Ludimar Hermann, Robert Meade Smith

Experimental Pharmacology

Ludimar Hermann, Robert Meade Smith

Experimental Pharmacology

ISBN/EAN: 9783337337964

Printed in Europe, USA, Canada, Australia, Japan

Cover: Foto ©berggeist007 / pixelio.de

More available books at **www.hansebooks.com**

EXPERIMENTAL PHARMACOLOGY.

A HAND-BOOK

OF

METHODS FOR STUDYING THE PHYSIOLOGICAL ACTIONS OF DRUGS.

BY

L. HERMANN,

PROFESSOR OF PHYSIOLOGY IN THE UNIVERSITY OF ZÜRICH.

TRANSLATED, WITH THE AUTHOR'S PERMISSION, WITH NOTES
AND ADDITIONS,

BY

ROBERT MEADE SMITH, M.D.,

DEMONSTRATOR OF PHYSIOLOGY IN THE UNIVERSITY OF PENNSYLVANIA.

WITH THIRTY-TWO ILLUSTRATIONS ON WOOD.

PHILADELPHIA:

HENRY C. LEA'S SON & CO.

1883.

COLLINS, PRINTER.

TRANSLATOR'S PREFACE.

THE translation of Hermann's Manual of Pharmacology was undertaken to furnish the student with a work that would assist him in his studies of the physiological action of drugs, enabling him to make the experiments himself that would otherwise require the assistance of the instructor.

The translator has attempted to elucidate the text with a careful selection of illustrations; and he trusts that his voluminous additions, which constitute nearly one-half of the entire volume, will render the work a more perfect guide to the student.

FEBRUARY 1, 1883.

1*

TABLE OF CONTENTS.

PART II.

INVESTIGATION OF THE GENERAL ACTION OF POISONS.

LIST OF·ILLUSTRATIONS.

INTRODUCTION.

THOSE substances are called poisons which, when introduced into the animal economy, produce disturbances of its normal functions. Occasionally the term is restricted to substances which are active in minute quantity, and when the disturbance produced by their action threatens the life of the organism. But since these distinctions are merely relative, they are superfluous, especially since the scope of toxicology can be defined by other considerations than by a definition of the word poison; the latter being only important from a medico-legal point of view.

Pharmacology in its widest scope embraces the study of drugs from all possible points of view, and the information thereby acquired may be useful under the most diverse conditions;—to the physician, to enable the recognition and proper treatment of cases of poisoning, or to permit of the use of drugs for therapeutic purposes;—to the public, to permit the avoidance of noxious substances;—to the physiologist and pathologist, to enable the application of information derived from the study of the action of poisons to the advancement of their sciences. The study of pharmacology can therefore be limited according as one or more of these points of view occupy the first place in the mind of the investigator. The public desires to know only what substances are poisonous, that they may be avoided, while their *modus operandi* is a matter of indifference. Those poisons which are suitable for use at the bedside will prove most interesting to clinicians.

Pure pharmacology is best advanced by the avoidance of

2

any special stand-point, in order that *all* of its bearings
may be equally appreciated, and still more, since the
advancement of pure science is always retarded by a
search for that only which promises immediately *practical* results. The history of the progress of the sciences
teaches that nearly all the most important discoveries,
even those subsequently of the greatest practical value,
resulted from investigations untrammelled by a continuous mindfulness of the merely practical. Thus physiology has rendered such inestimable assistance to the
progress of practical medicine that she can well be regarded as her handmaid ; but, nevertheless, physiology is
a pure science, which, like physics and chemistry, should
be studied for its own worth, without being hampered by
doubts as to whether its results are immediately applicable
to practical medicine or not. So also pharmacology is
growing more and more worthy of occupying a similar
position, though it must be acknowledged that, as yet, it
is not bounded by such sharply drawn lines as to constitute a distinct science. Much, however, can be gained in
this direction by constantly bearing in mind that pharmacology has for its object the recognition and study of all
changes which a foreign body can undergo or produce,
otherwise than traumatically, in the organism, while the
questions as to whether the substance under study can
ever be likely to prove a poison to man, or whether it has
properties which warrant its use as a medicament, should
be kept in the background.

Consequently every substance which possesses *any*
active properties should prove of interest to the investigator in the domain of pharmacology; while naturally
those substances will be preferred which are either quite
unknown, which show results entirely novel, or whose
action admits of predetermination from a theoretical
point of view, as from the stand-point of chemical composition. And it should, moreover, be remembered, that
even substances which themselves evoke no symptoms in
the organism, may form worthy subjects of pharmacological investigation as throwing possible light, in the

changes which they undergo in the system, on the behavior of other more active poisons.

The object of pharmacology, therefore, is to acquire
familiarity with the peculiarities and actions of poisons,
to carefully analyze all processes which they evoke so as
to obtain a complete picture of their mode of action.
This object cannot be attained by mere observation of
cases of poisoning in man, although such cases, when
properly studied, may be of the greatest service; since
it is often only by such means that we are enabled to form
conclusions as to the action of the drug on man.

Experiment must be the instrument most relied on in
pharmacology; not only because it alone permits the
study of all poisons in all doses, and on the most various
organisms; but because it is indispensable to the acquire-
ment of any more profound knowledge of the *modus
operandi* of poisons than can be obtained by a mere inspection of cases of poisoning.

Observation of cases of poisoning only furnishes a
coarse method of observing the prominent features of the
action of the poison, while experimentation alone renders possible that analysis of the information so acquired
which enables the deduction of an opinion as to the
changes which the poison itself undergoes, the means by
which those changes are produced, and their results and
the special action which they may exert on individual
organs. The extent to which such deductions will become possible will depend upon the perfection of the
experimental art and the state of our knowledge of the
normal functions.

Experimentation on living animals is chiefly employed
by physiologists, who are consequently pre-eminently
suited for the study of pharmacology. But this is not
the only explanation of the fact that the authors of
most of our most valuable papers on the action of drugs
are practical physiologists; a deeper-lying reason for the
devotion of physiologists to pharmacology is to be found
in the fact that nearly every addition to our knowledge
of the action of a poison marks at the same time an ad-

vance in our knowledge of the normal organism, and can hence be regarded as a step in the development of the science of physiology.   Physiologists, therefore, rightly regard pharmacological investigation as one of the most important modes of advancing their science.   Occasionally, also, pharmacology furnishes an instrument of experimentation of the most delicate character ; as an illustration of this we have only to mention curare, whose employment in pure physiology has been most fruitful of valuable results.[1]

---

[1] [In this connection see Bernard's valuable paper on "Les poisons comme méthode de vivisection," in the Revue Scientifique, 1875.]

METHODS

OF

PHARMACOLOGICAL INVESTIGATION.

The first indication as to the poisonous action of any substance is usually to be found in reports of cases of poisoning occurring in man. But the histories of such cases, though they may indicate the most promising line of study, in the majority of instances give us only an imperfect, and in the light of experiment, often an inaccurate picture of the action of the poison. The consideration of suggestions derived from such sources, though they may often facilitate the attainment of definite results, just as a qualitative assists a quantitative chemical analysis, will, however, for the present be deferred.

When it is desired to determine the mode of action of a substance of whose *modus operandi* no conception has been formed, one of two lines of investigation may be followed: Either an animal is suitably brought under its influence and a general picture of its working obtained, which, imperfect though it be, is far better than the generality of reports of cases of poisoning, and which may serve as a starting point for the explanation, by subsequent experiments, of individual symptoms; or the action of the drug on separate, isolated organs, may first be studied, and, assisted by this preparatory knowledge, its influence on the animal system then determined.

In general, the first method leads most directly to the desired end. Nevertheless, we will here follow the latter

2*

plan, since the analysis of results obtained in the animal organism presupposes a certain amount of experience in explaining elemental disturbances ; this the beginner can best acquire by the method first indicated. We will therefore commence with the methods for the examination of the action of poisons on isolated organs.

PART I.

STUDY OF THE ACTION OF A POISON ON ISOLATED ORGANS.

————

THE organs of the cold-blooded vertebrates, especially the frog, are best suited for this method of pharmacological investigation, since, as a rule, they are equally with those of mammals susceptible to the action of poisons, and they may be isolated from the circulation for quite a while without undergoing any essential alteration. Not only excised organs, such as the heart, muscles, nerves, etc., can be used for this purpose, but the action of the drug can even be restricted to certain portions of the economy while still in the body; since, on the one hand, their exposure does not necessarily entail any general disturbance of function, and, on the other, the exclusion from the circulation of certain parts of the body neither destroys the functions of the isolated parts nor interferes with the normal condition of the remainder. For example, the action of the poison can be limited to one extremity; or the entire body, with the exception of one extremity, can be exposed to its influence.

Until quite recently, the only organ of warm-blooded animals which was capable of isolation for pharmacological studies was the blood, whose physiological status, especially when kept at the normal temperature, is readily maintained outside of the body.

[By improved methods of research this line of study on mammals can now be greatly extended. The methods will be given in their appropriate chapters.]

SECTION I.—**Action on the Blood.**

Blood in considerable quantities can be readily obtained only from warm-blooded animals, the selection of the animal depending on the special point to be studied: thus, when alterations in the hæmoglobin are to be examined, it is advisable to employ easily crystallizing blood, such as that of the horse, dog, or guinea-pig.

[In order to collect blood, either arterial or venous, uncontaminated with foreign matters, it is necessary to isolate the artery or vein and insert a canula into the vessel. Various forms of canulæ may be employed. The simplest, and therefore the best, is readily prepared by drawing out in a lamp or gas-flame, a piece of narrow glass tubing until the desired diameter is attained; when by further heating the points of junction of the narrow portion with the remainder of the tube, and gently drawing out the tube, a constriction is made at these points, a and b (Fig. 1, D). The narrow portion of the tube

Fig. 1.

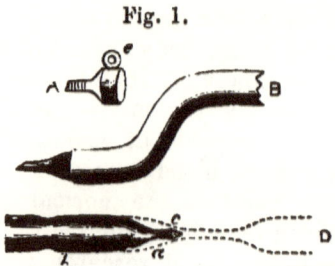

A is the ordinary form of metal canula, with a ring at *e* by which it can be tied to any larger tube. *B* is a holder made of a metal tube with a wooden point to facilitate the introduction of *A*. *D* represents the manner of making glass canulæ.

is then gently heated over a small flame, drawn out and then filed across at *c*. The point of the canula thus made is then to be ground down obliquely by rubbing on a hone and the edges rounded in the flame; the large end of the canula is now tied into a short piece of rubber tubing and is ready for use. In this way a number of canulæ, which are easily cleaned and inserted, can be made in a few minutes.

In order to insert a canula into a vessel the surface must first be freed from hair, and after narcotization, when permissible, the skin divided by an incision about an inch long in the line of the vessel. The connective tissue and subcutaneous muscles over the vessel may then be torn away with a pair of blunt hooks, or two pairs of forceps, and the vessel carefully and thoroughly freed from its connective tissue sheath; on the success of this step, particularly in the case of veins, will depend the readiness with which the canula can be inserted, since after the vein has been opened it will immediately collapse and it will then be found much easier to insert the canula into the space between the sheath and vessel than into its proper place. Ordinarily, except in the case of large animals with strong connective tissue fibres, this method of exposing the vessel is preferable, in the avoidance of hemorrhage, to the use of any cutting instrument; when, however, knife or scissors must be employed, the bleeding, if any is caused, must be controlled by ligatures or torsion. After the vessel has been exposed, a pair of forceps can be passed under it and then opened, thus serving to maintain the vessel in a position where it can be readily operated on. Three ligatures are then passed under the vessel; if it is an artery, the one farthest removed from the heart is first tightly tied, so as to occlude the vessel, and the one nearest the heart is then tied in a slip-knot so as to be readily removed. A snip is then made in the vessel midway between the two ligatures with a pair of scissors which cut well at their points, and the canula, which may conveniently be held on a piece of wood, then inserted towards the heart and bound fast by the middle thread, the knot being tied around the neck of the canula. The canula is further prevented from slipping out by bringing the ends of the middle ligature parallel with the tube and then encircling them with a thread tied around the large end of the tube, and tying the ends of both sets together.   The same manipulations are used when it is desired to insert a canula into a vein, with the exception that the proximal ligature is first tied so as to distend the vessel, and the slip-knot then tied with the distal ligature, the canula pointing towards the periphery instead of towards the heart, as in the case of the artery.   Instead of slip-knots, spring-clips may be used to compress the vessels.   Blood is then collected by connecting the canula with a clean glass tube leading to a well-cleaned vessel of any sort, preferably glass, and then untying the slip-knot, or removing the bull-dog forceps or clip.

In order to collect blood free from access of air, the canula may be connected by tubing with the upper end of a burette, protected by a stopcock, the lower end of which communicates, a stopcock intervening, with a movable vessel containing mercury.   The mercury reservoir is first to be raised to such a height that the mercury rises to the top of the burette and commences to flow over ; the lower stopcock is then closed and the reservoir depressed ; the clip or knot is then removed from the vessel and the instant the blood reaches the end of the tubing connected with the canula, all air being expelled, the end of the tube is slipped over the top of the burette and bound fast.   On now opening the lower stopcock of the burette the mercury falls and draws after it into the burette the blood from the vessel.   As soon as enough is obtained, the stopcocks are closed, and the tube may be shaken to defibrinate its contents.]

In addition to the action of the drug on blood removed from the body, it is often advisable to examine the blood while still circulating in the vessels, as can be readily done in the tongue, or swimming bladder of frogs, or in the mesentery of mammals, with the aid of the microscope ; in such experiments the poison is injected into the general circulation.

As far as pharmacological studies have yet taught, we may have to deal with the following forms of alteration of the blood produced by poisons.   We commence with those most readily detected.

1. ALTERATIONS IN CONSISTENCE, FROM ACTION ON THE ALBUMINOIDS OF THE BLOOD.—Caustic alkalies can convert the blood into a gelatinous mass by the production of alkali-albuminate. By prolonged action, alcohol coagulates the albuminoids of the blood; many metals, alcohol by short action, aniline, etc., precipitate them. Precipitation of globulin is produced by acids, and when in concentrated solutions, other albuminoids also are thrown down. Alterations in color ordinarily accompany changes in consistency; these will be subsequently studied.

As a rule it will not be possible to determine the character of the precipitate; at least when some general idea of the action of the substance experimented with is not already possessed, no general rule for its closer study can be given, since so little precise knowledge is possessed of the alterations of the albuminoids, that each case requires a special chemical study. It may be recommended, however, in order to obtain some conception of the nature of the albuminoid to whose alteration the changes are due, not only to experiment with blood collected directly from the artery or vein, but also to examine into the action of the poison on defibrinated blood and on blood serum.

2. ALTERATIONS IN THE ALKALINE REACTION OF THE BLOOD.—The reaction of the blood cannot be directly tested with litmus paper. One of the simplest methods is that recommended by Kühne.[1] It consists in placing the blood in a small dialyzer, made by moulding a piece of parchment paper into the form of a minute cup, and floating the dialyzer on the surface of a little distilled water in a watch glass. A little blood is then placed in the dialyzer thus constructed and after a short interval the reaction of the colorless drop of water in the watch-glass is tested.

[This method is, however, not perfectly reliable, since

---

[1] Archiv f. Path. Anat., xxxiii. 95.

the blood will probably coagulate before the reaction can be determined, and the alkalinity may possibly be thereby altered; but, though not giving, strictly speaking, the reaction of fresh blood, it will generally serve the purpose.

Liebreich recommends the use of a slab of neutral plaster of Paris stained with neutral litmus solution; a few drops of blood are allowed to fall on the slab, and, after a few seconds washed off with a little water: the blue coloration can then be detected where the blood rested.   In order to determine the degree of alkalinity, a standard solution of tartaric acid may be made by dissolving 7.5 grm. of crystallized tartaric acid in a litre of distilled water; 1 c. c. of this solution should exactly neutralize 0.004 grm. of sodium hydrate.   The acid solution is added from a burette to 50 or 100 c. c. of the serum or blood, a drop of the mixture being placed from time to time upon the slab colored with litmus; the addition of acid is continued until the reaction is faintly acid.   The alkalinity of the blood may then be expressed as corresponding to x milligrammes of sodium hydrate per 100 c. c. of blood.][1]

An even shorter method [based on the fact that blood coloring matter does not diffuse out of the blood corpuscles into strong solutions of common salt] is that advised by Zuntz.[2]   A drop of blood is placed on a strip of glazed litmus paper which has been previously moistened with a concentrated solution of sodium chloride, and, after a few minutes' contact, drawn up with a pipette or filter paper.   It is only necessary to be sure that the salt solution, which itself often becomes alkaline when kept for any length of time in glass vessels, is neutral. By this process it is also possible to determine the degree of alkalinity of the blood by adding standard solutions of acid to the blood and employing this method to determine when the reaction is acid.   [It may, how-

---

[1] Gamgee, Physiological Chemistry, vol. i. p. 177.
[2] Centralb. f. d. Med. Wissen., 1867, 531.

ever, be objected to this method, that the addition of the salt causes alterations in the corpuscles, producing shrinkage and osmoses into the blood plasma. Schäfer[1] recommends the use of the delicately colored glazed English litmus paper; all that is necessary, according to him, is to place a drop of blood on the paper and after a few seconds to wipe it off. The blue patch will indicate the alkalinity, since the alkaline salts will soak into this paper faster then will the hæmoglobin.]

3. ALTERATIONS IN THE RED BLOOD-CORPUSCLES.—For studies on this point, when the poison is a liquid and not volatile, a few drops of perfectly fresh, defibrinated blood are placed on an object-glass, covered with a cover-slip, and a drop of the poison placed on the slide and allowed to mingle with the blood under the cover-slip. If it is not desired to study all the steps in the process of change, if any should be produced by the drug in question, the poison and the blood may be first mixed and a drop of the mixture then placed upon a slide and examined under the microscope. If, however, the drug is in the form of a solution, great care must be observed in attributing the results to the action of the drug, as there is scarcely any known solvent which does not itself produce change in the red blood-corpuscles; in such cases control experiments must be made with the solvent alone. In suitable cases, 0.5–0.7 per cent. solution of sodium chloride, which itself produces no sensible action on the corpuscles, may be used as a solvent. If the drug is a volatile liquid, the blood may be subjected to the action of its vapor by passing air or any indifferent gas through the liquid and then through the blood, in the manner to be described under the study of the action of gases [or any of the gas-chambers, e. g., Stricker's may be used].

The simplest method of subjecting blood to the action of gases or vapors, is to pass them through a tube reaching

[1] Journal of Physiology, Jan. 1882.

3

to the bottom of the vessel in which the blood is contained. The process may be accelerated by closing the vessel which contains the blood after the air has been displaced by the gas, and then agitating. If it is desired, during the action of the gas, entirely to prevent the access of air, a small quantity of blood may be passed up into an ordinary Torricellian barometer tube, and then the gas conducted through it; the action of the gas may then be facilitated by closing the open end of the tube with the finger and then shaking, though it should be remembered that shaking blood with finely divided solid particles, and therefore probably also with the globules of mercury, may itself cause an alteration in the blood-corpuscles.[1]

The continuous passage of gases through undiluted blood, as is here desirable, is, by any of the methods yet mentioned, an extremely unsatisfactory and incomplete procedure on account of the foam produced; since each gas bubble only comes into partial contact with a limited portion of the blood, and the complete result of the treatment may only appear after a long interval of time. By the following method,[2] however, the object is rapidly attained; a vertical glass tube, about 5 mm. in diameter, and on which there may with advantage be blown a number of bulbs, is to be bent at its lower end so as to form a short arm forming an acute angle with the vertical portion of the tube. A small portion of blood is now drawn by suction into the tube so as to occupy a part of both of its arms and the shorter arm is then connected with the gas generator. The gas then forces the blood up into the vertical arm, bursts through it in bubbles, and the blood flows back along the sides of the tube to be again forced up by the gas, thus always bringing fresh surfaces into contact with the gas, and producing a complete action in the shortest possible time.

In certain special cases, it may be desirable to keep

[1] Compare A. Rollet, Sitzungsber. d. Wiener Acad. Math. Natur. Wissen. Cl. 2 Abth. lii. 246.
[2] Hermann, Arch. f. Anat. u. Phys. 1865, 471.

the blood, which has been subjected to the action of the
gas, free from access of air; it is then only necessary to
seal the tube, after introducing the blood and gas, by
fusion at two points where it has previously been drawn
out to a thread. If it is necessary to observe the action
of the gas at a high or low temperature, the tube may
be bent in the form of a $\vee$, and during the experiment
immersed in a water-bath or cooling-mixture.

The methods as yet described do not permit of continuous observation from the commencement of the action of
the gas on the blood-corpuscles; this, however, is rendered possible by the use of the different forms of gas-
chambers for microscopic use, most of which also permit
of the object being studied at different temperatures.[1]

The alterations of the red blood-corpuscles produced by
different agents,[2] as far as are yet known, may be changes
in shape or color, or complete decolorization or solution.
The latter two processes, of which the first may, for example, be produced by water, and the second by ether,
cause the blood to assume the appearance known as
"laky." While normal blood, even in the thinnest
layers, is perfectly opaque, and when allowed to flow
down the side of a glass forms irregular streaks, after
decolorization, or solution of the red corpuscles, it becomes transparent and perfectly homogeneous, like red
varnish. Under the microscope, in both cases, the serum
is seen to be colored, and the corpuscles are either entirely
invisible, or in simple decolorization, occur in the form of
perfectly pale, scarcely perceptible spheres. ["Stro-
mata" according to Rollet.]

The change of the disk-shape into the spherical form
seems always to precede the loss of color, even when
the latter is caused by the total solution of the corpus-
cle. In addition to this spherical alteration in the
shape of the red blood-corpuscles, they may also take on

[1] Kühne, Arch. f. Path. Anat. xxxiv. 423, and Stricker's Hand-
buch, p. 411.
[2] The alterations produced by heat, cold, electricity, etc., do not
fall within the province of this work.

a jagged, contracted form (crenated) from the action of certain reagents, for example, as occurs on the addition of concentrated salt solution or on drying.

Alterations in color of the corpuscles will be mentioned under the heading of changes in the coloring matter of the blood.

4. ALTERATIONS IN THE WHITE CORPUSCLES.—The white blood-corpuscles may be studied in precisely the same manner as the red. In the former case, however, it is necessary in addition to examine into the effect of the agent on the contractility of the white corpuscles : this may be accomplished by experimenting with a drop of blood on the warm stage of the microscope.

5. ALTERATIONS IN THE COAGULABILITY OF THE BLOOD.
—In order to form any conclusion as to this point, it is evidently necessary to subject the blood to the action of the poison the instant that it leaves the bloodvessels. Co-agulation may be either accelerated, retarded, or pre-vented.

Should the blood, after the addition of the poison, appear to coagulate more rapidly than normal, a control experiment must be made by adding some of the poison to defibrinated blood in order to determine whether the coagulation is due to fibrin-formation or to the coagu-lation of albuminoids, etc. [The coagulum formed, and the process of its formation, may also advantageously be examined under the microscope.]

According to the researches of Alexander Schmidt,[1] retardation or prevention of coagulation depends upon action of the poison either on the fibrinogen or the fibrino-plastin.

When, therefore, coagulation is retarded, the mode of action may be determined by allowing the drug to act either on isolated fibrinogen or fibrinoplastic substance, and then determining their individual activities. Fibrinogen

---

[1] Arch. f. Anat. u. Phys. 1861, p. 545, 675 ; 1862, 428, 533.

can be readily obtained in the pericardial fluids of man or other mammals, and in most hydrocele fluids; its presence, however, in these liquids must be first established by the production of a coagulum when they are added to the serum obtained by subjecting blood-clots to pressure. Fibrinoplastic substance may be obtained from the fluid thus obtained from blood coagula. Fuller details for investigations on these points must be obtained from Schmidt's memoir.

[The comparatively recent researches of Hammarsten[1] on the chemistry of blood coagulation have rendered necessary some modification of the theories of Schmidt and Buchanan. According to his investigations the evidence is decidly in favor of the view that fibrin is produced by the decomposition or change of fibrinogen, and that this change is connected with the presence of a ferment derived from the breaking down of the white corpuscles, or, perhaps, from the granular bodies recently described by Bizzozero. Paraglobulin, or the fibrinoplastic substance, is therefore only indirectly, if at all, connected with the process; at any rate the idea that fibrin can only be formed by the union of fibrinogen and fibrinoplastin in the presence of a ferment is no longer tenable.

The same methods, however, can be employed as indicated above, since the ferment is contained in the serum expressed from blood clots.]

6. ALTERATIONS IN THE COLORING MATTER OF THE BLOOD.—For studies on this point it is advisable to dilute the blood which has been treated with the poison, with water, so as to be able to study changes in its color with transmitted light; generally the dilution should precede the addition of the poison.

Changes in hæmaglobin are best and easiest studied by means of the spectroscope, as described in Hoppe-

---

[1] Pflüger's Archiv, xiv. xvii. xviii. xix.

Seyler's Chemische Analyse. [See also the paper by Gamgee on the action of " Nitrites on the Blood," Proc. of the Roy. Soc., 1868, p. 389.]

7. ALTERATIONS IN THE GASES OF THE BLOOD AND IN THE POWER OF ABSORBING GASES.—From the fact that when blood is drawn from a vessel changes in the proportion of its contained gases at once commence, the action of a drug on the blood gases can be only partially determined.

For example, if oxygen is diminished in proportion, or entirely absent, it would be made evident by the change of color to a dark venous hue and by the characteristic spectroscopic changes ; changes in percentage of carbonic anhydride are however much more difficult to establish.

Accurate experiments on either of these questions are extremely difficult ; the blood must be collected without access of air and then subjected to the action of the poison and the gases then determined ; or its power of absorption for certain gases established after the blood, from which all gases have been exhausted, has been brought under the influence of the poison. Both these processes are complicated and will be rarely requisite in pharmacological studies : they will therefore not be entered into here.

8. ALTERATIONS IN THE OZONE AND OZONIZING POWER OF THE BLOOD.—For the determination of the first of these points, the blood must be preserved throughout free from air, since ozone is formed as soon as blood comes in contact with the atmosphere. The blood must therefore be collected in a vacuum, or in an atmosphere free from oxygen, then mixed with the poison, and the surrounding air, which is then allowed access, tested with freshly prepared guaiacum test-paper (or strips of paper moistened with a solution of iodide of potassium and starch ; if ozone is formed it will unite with the potassium, the iodine will be set free and color the starch blue).

Full details for such studies will be found in Kühne and Scholz on " Ozone in the Blood."[1]

In order to determine the ozonizing power of the blood, it must be treated with carbon monoxide gas, exhausted of oxygen, and then brought under the action of the poison; the presence or absence of ozone in the air can then be determined by test papers. In order to determine the readiness with which the blood yields up its ozone (?) a more delicate test than guaiac must be employed; the best process is to add a small quantity of tincture of guaiacum to a drop of oil of turpentine and then a few drops of the blood which has been acted on by the poison. If the power of transferrence is unaltered, the fluid will immediately become dark blue.

For the study of other alterations of the blood, such as precipitation, or alterations in the poison produced by the action of the blood, no general rules can be given.

SECTION II.—**Action on the Muscles.**

Only a few poisons are suitable for direct application to excised muscles; of course in these cases, the muscles of cold-blooded animals alone can be used, those of the frog being nearly always selected.

Since most poisons must be used in the form of a solution, and since nearly all solvents, even distilled water, are irritating to muscles, the direct immersion of a muscle in a solution of a poison is always a doubtful experiment. A solution of common salt, 0.5–0.7 per cent., is the most neutral solvent that can be selected.[2]

A better method than the immersion of the muscle in the solution, is to determine the local action of the poison by injecting the solution into the bloodvessels; [this can be readily accomplished in the frog by inserting a canula in the bulbus aortæ, washing out the circu-

---

[1] Arch. f. Path. Anat., xxiii. p. 96.
[2] See Nasse, Arch. f. d. ges. Physiol. ii. 97.

lation with salt solution, and then making an injection of the poison. By this means the effects of local action on the muscles can be readily studied.]

Gases and vapors are best suited for direct application to the muscles; to accomplish this, the muscle is suspended in a bottle, or cylindrical vessel, closed by a cork through which pass two tubes; one simply traversing the cork, the other, through which the gas is to be conducted, passing to the bottom of the vessel and dipping under the surface of a layer of water. When the gas is forced through the vessel it comes in direct contact with the muscle and, at the same time, keeps the atmosphere around the muscle saturated with aqueous vapor, a condition absolutely necessary for all experiments on muscles or nerves.

It is convenient to so suspend the muscle, that it can be readily irritated while still within the vessel. A fine wire, passing through the cork and ending in a hook on which the tendon or bone is suspended, can serve as one electrode; a very light metal hook, connected with a fine spiral wire which also passes through the cork, can be inserted in the other tendon and serve for the second electrode. For experiments with gases and vapors the thinnest possible muscles, such as the *sartorius* of the frog, must be selected, since the gases act solely on the superficial fibres.

For the majority of poisons it is best to first administer the drug, in a way which will be later indicated, to uninjured animals, and then, after the action of the poison has become evident, to excise their muscles for investigation. It is very often desirable, as a control experiment, to study at the same time an unpoisoned muscle of the same animal. In the frog, which is alone suitable for such experiments, single groups of muscles may readily be protected from the poison by ligation of their arteries before the drug is administered. This is most easily accomplished by shutting off the blood supply from one entire limb by ligation of the common iliac artery, or the femoral artery may be ligated at the

upper part of the thigh and the foot and leg of that side
be thus cut off from the circulation.

[Ligation of the bloodvessels of a limb will not invariably succeed in preventing access of the poison to that
member, since Ringer and Murrell[1] have found that certain substances, such as potash salts, are capable of being diffused through tissues shut off from the circulation
almost as rapidly and as thoroughly as when their blood
supply has not been interfered with. In cases, therefore,
where ligation of the blood supply of a limb appears to
produce no modification in the character of the symptoms,
which on other grounds are presumably of peripheral
origin, the tissues should always be examined for the
presence of the poison.]

The ligation of a common iliac artery is performed in
the following manner: The frog being fastened on his
belly, in the lower part of the back three bony lines are
easily detected, the two iliac bones and between them
the coccyx; an incision is to be made, parallel to these
bones, between the ilium and coccyx, through the oblique
muscular mass, the ileo-coccygeus; the iliac plexus can
then be seen, composed of the prominent nerve trunks,
lying deep within the peritoneal cavity, and at its inner
border, the common iliac artery. The artery can then
be readily isolated with a delicate blunt hook, and a fine
thread passed around it and ligated.

To ligate the femoral artery, a frog is fastened in the
same manner and a longitudinal incision, _m_, _n_ (Fig. 2.),
made through the skin in about the middle of the posterior
surface of the thigh; a deep longitudinal furrow is then
seen lying between the _vastus externus_ on the outer side
and the _semimembranosus_ on the inner side, with the thin
round biceps lying in the furrow (Fig. 3); on drawing
the vastus with the biceps towards the outer side and cutting through the fascia, the sciatic nerve and femoral
artery are seen running together, the latter being easily
recognized by its black pigmentation; it is to be sepa-

[1] Journ. of Phys., vol. i. pt. i.

rated from the nerve with a blunt hook and ligated. If it is desired to ligate the artery near the knee-joint, the

Fig. 2.

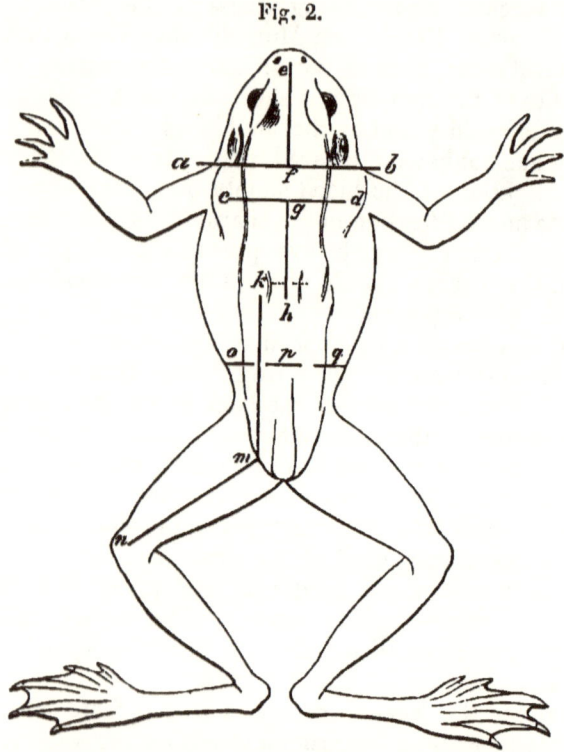

Diagram of a frog to show the lines of various incisions.

biceps is first to be divided at its lower end and turned upwards; the artery can then be readily reached.

The method of preparation of the muscle for study after poisoning, will depend upon the special point to be investigated. For the examination of the relative irritability, etc., it is always advisable to isolate the muscle with as long a piece of its nerve as possible ; for most purposes, especially when it is desired to load the suspended muscle, the gastrocnemius in connection with

the sciatic nerve and femur is the most suitable preparation. Ordinarily, however, the condition of irritability
is alone dealt with, and for the examination of this point
the method of Du Bois Reymond is perhaps the simplest.
His preparation is made as follows : After destruction of

Fig. 3.

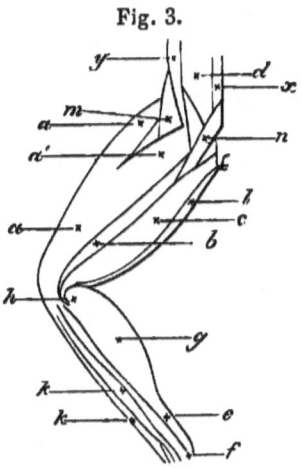

Diagram of the muscles of the leg of a frog, posterior surface. *a*, triceps
femoris ; *b*, biceps femoris ; *c*, semimembranosus ; *d*, coccygeo-iliacus ; *e, f,*
tendo Achillis ; *g*, gastrocnemius ; *h*, head of gastrocnemius ; *k*, peroneus, the
outer muscle marked *k* is the tibialis anticus ; *l*, rectus internus ; *n*, pyriformis ;
x, coccyx ; *y*, ilium ; *a'*, vastus externus.

the central nervous system, by thrusting a wire down
the spinal canal and into the cerebral cavity, the frog is
cut in two in the lower dorsal region with a pair of strong
scissors, and the skin stripped off the lower extremities ;
the furrow between the vastus externus and semimembranosus is then sought for, and the sciatic nerve isolated
as near the knee as possible. A blade of a sharp pair
of scissors is then carefully passed under the nerve and
the leg amputated near the knee, care being taken not
to injure the nerve. The foot is then held between the
fingers, and the nerve carefully dissected out from below
upwards to the spinal column, and all branches divided,

great care being taken not to stretch the nerve or injure it with the scissors. The most difficult part to isolate safely is where the nerve enters the abdominal cavity. When the spinal column is reached, the plexus which forms the sciatic is to be divided. [The preparation generally known as the "nerve-muscle preparation" is obtained by a slight modification of this method. The femur is divided about half an inch above the knee and the nerve prepared as above; the gastrocnemius muscle is then exposed, its tendon divided as near the foot as possible, and the muscle separated by a blunt probe from the rest of the leg which is then divided *below* the knee. The preparation then consists of the entire sciatic nerve from the knee up to the vertebral column, the knee-joint with a portion of femur by which it can be suspended, and the gastrocmius muscle (Fig 4).]

Fig. 4.

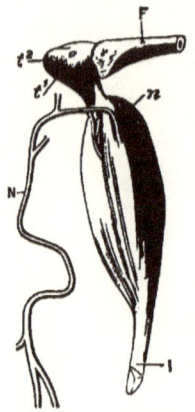

The nerve-muscle preparation. F, end of femur; N, sciatic nerve; I, tendo achillis; t', origin of lesser tendon of gastrocnemius.

All preparations of frogs' nerves and muscles must be carefully protected from drying.

1. Measurements of the electro-motive power of poisoned muscles will be seldom required, since there seems to be ground for believing that the electro-motive power runs parallel with its irritability. Should, however, it be desired to examine into this question, the methods may be obtained in Du Bois Reymond's writings.

2. The alterations which are generally met with in muscles subjected to the action of poisons are changes in irritability, occurring in various degrees.

Muscular irritability may be estimated either by the direct or the indirect method. In general, the presence of muscular excitability, as evidenced by indirect stimulation (*i. e.* by nerve irritation) presupposes the existence of direct excitability, although it is conceivable that muscles may be susceptible only to stimuli coming through the nerves. Examination of indirect muscular excitability includes, therefore, the examination at the same time of nerve excitability.

1. EXAMINATION OF INDIRECT MUSCULAR IRRITA-BILITY.—The general rule for this class of experiments is that two muscle preparations, a poisoned and non-poisoned, must be continually compared in order to elimi-nate changes in irritability produced by the drug from natural changes, such as would accompany the death of the nerve, etc. An exception to this rule is only permis-sible when the changes produced by the drug are of the most marked character, such as the production of entire loss of irritability.

Both preparations must be taken from the same animal and should be corresponding animal parts, one having been previously protected from action of the poison by ligation of its bloodvessels. The method has already been given. Convulsions or fibrillar contractions caused by the operative procedure, must be allowed to pass off be-fore the changes in irritability, if such exist, are mea-sured.

Unless there is some indication for the employment of a special form of stimulation [such as mechanical, ther-mal, or chemical], electricity is always preferable, and the most convenient form is the tetanizing induction cur-rent.

[For the ordinary application of the induced tetanizing current the most convenient apparatus is the Du Bois Reymond induction coil. This instrument, as arranged for use, is seen in Fig. 5. The positive pole, x, of the bat-tery is connected by a wire with the binding post a, and the negative wire, y, with the post d; the current then passes through the vibrator b to the screw c, and then around the primary coil p, to the electro-magnets $m\,m$, the post d, and so back to the battery (Fig. 5). As soon as the current enters the primary coil it causes an instan-taneous induced (closing) shock in the secondary coil (not

4

seen in the figure) ; when the current passes around the spirals *m m*, it renders their soft iron cores magnetic, and therefore draws down the hammer *e* of the vibrator *b*, and interrupts the passage of the current through the pri-

Fig. 5.

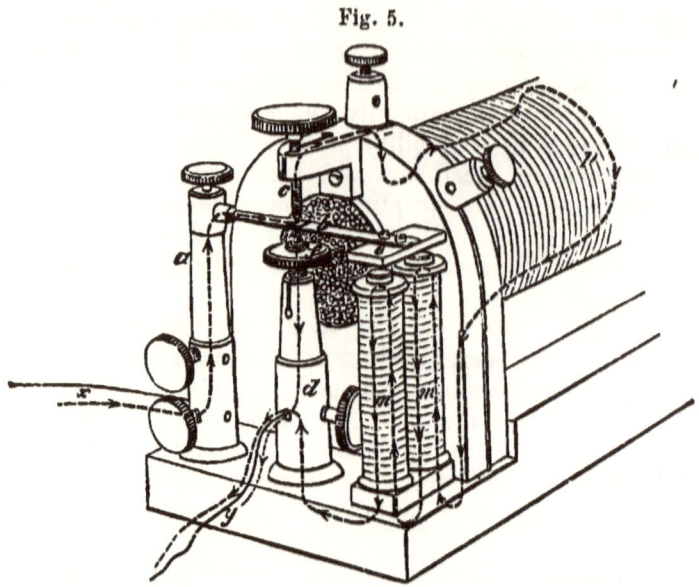

Du Bois Reymond induction apparatus.   (From Foster's Physiology.)   For explanation, see text.

mary coil, by breaking the contact between *b* and *c*, and so causes an instantaneous induced current (breaking shock) in the secondary spiral.   As soon, however, as the current is so interrupted, the cores of *m m* lose their magnetism, the hammer flies up again, makes contact with *c*, so allowing the current to pass, is again interrupted, and so on.   The binding posts on the secondary coil are connected with wires to the electrodes.

Ordinarily it is sufficient to compare the strengths of current necessary to produce contractions in the poisoned and normal muscles.   This can be done by first ligating the bloodvessels of one hind leg and then injecting the

poison into the abdominal lymph sac; the two sciatic
nerves (or gastrocnemii muscles when direct stimulation
is being employed) are exposed and divided high up;
their ends are then placed on two pairs of electrodes and
connected, a double key (Fig. 6) intervening, with the

Fig. 6.

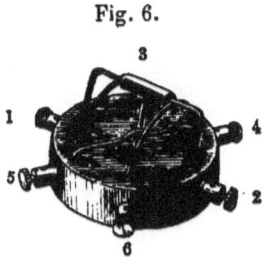

Pohl's commutator or double key. (From Sanderson's Handbook.) To use
this instrument as a double key, the diagonal wires connecting 3 and 6, and 4
and 5, are first removed; the binding screws 1 and 2 are then connected with the
secondary coil of the induction apparatus, and 3 and 4, and 5 and 6, each with
a set of electrodes. As represented in the figure, the posts 2 and 4, and 1 and 3,
are electrically connected; when the cradle is rocked so that its ends dip in the
mercury in the cups at 5 and 6, the electrodes connecting with 3 and 4 are cut
off, and 5 and 6 receive the current.

In order to use this instrument as a current reverser, the centre wires are
placed in position as shown in the figure. The + pole of the battery is connected with 2 and the − with 1; 4 consequently is the + electrode and 3 the
− electrode. When the cradle is rocked so as to dip into 5 and 6, the + current
passes through the cradle into 6, then across the connecting wire to 3, which
thus becomes the + electrode; the − current follows an analogous course.

secondary coil of Du Bois Reymond's induction apparatus.
The secondary coil is then pushed some distance away
from the primary coil (the greater the distance between
the coils the weaker the stimulus), and the current
closed in the primary coil by means of a Du Bois Reymond key (Fig. 7); if no contraction is caused in the
muscle connected with the apparatus, the secondary coil
is pushed up towards the primary until a contraction is
evoked, and the time and the distance between the two coils
recorded; the cradle of the double key is then reversed
so as to send the current through the other electrodes,
and the weakest current which will cause a contraction

determined in the same manner. It may also, occasionally, be well to determine the irritability of the nerves before the drug is administered, and then again at intervals afterwards.]

Fig. 7.

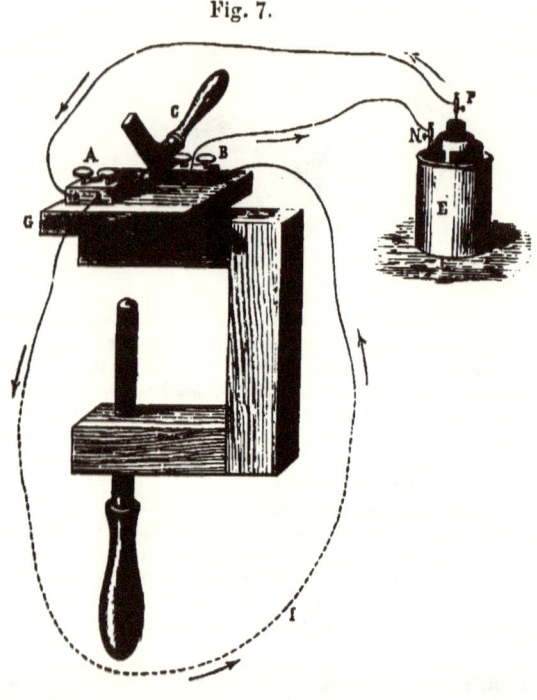

Du Bois Reymond key. (From Sanderson's Handbook.) When the wires are arranged as represented in the figure, the current is closed in the primary coil by opening the key, because when closed the key, offering so much less resistance than the coil of the induction machine, serves to short-circuit the current.

If it be desired to make very accurate and precise studies as to the condition of irritability, especially as to the influence exerted by the direction of the current and by opening and closing shocks, another form of apparatus must be used. As a stimulus, a constant galvanic current is to be employed, and the strength of the current graduated by a rheochord; the opening and closing of the circuit may be produced by an ordinary key or by dipping one end of the conducting wire into a cup of mercury in which the other end of the wire is fastened, so making a closing shock; when the wire is lifted out from the mercury, it of course being understood that the nerve or muscle is included in the circuit, an opening or breaking shock is made. The direction of the current may be altered by any of the various forms of commutators, of which Pohl's is the simplest. Unpolarizable electrodes should always be used;[1] while the height of contraction can be measured by means of Pflüger's myographion.

[The arrangement of apparatus represented in Fig. 8 may be used in such experiments. The induction coil is so arranged as to allow of either single opening or closing induction shocks being employed. When the key F is depressed, the current passes through the primary coil by the two binding posts on the top of the instrument, without passing through the vibrator, and so produces a single instantaneous closing shock in the secondary coil. When the key F is elevated, the current ceases to pass through the primary coil and so produces an instantaneous breaking induction shock. The wires from the secondary coil are short circuited by the Du Bois Reymond key C, so that when closed the key offers so much less resistance than the wires and parts beyond that no current reaches the electrodes. If, therefore, it is desired to pass a single closing shock through the nerve or muscle, the key C is first opened and then the key F closed, and the key C is then closed before the key F is opened; if a single breaking shock is desired, the key C is first closed and then the key F; and while the latter is closed the former is opened, and then the current through the primary coil broken by elevating F. A represents a convenient form of moist chamber in

---

[1] Rosenthal, Electricitätslehre, 2 Aufl. 57. A simpler form is described by Hermann, Archiv f. d. Gesamt. Physiol. iv. 211.

Fig. 8.

Arrangement of apparatus for experiments on muscle and nerve. (From Foster's Physiology.)   The nerve-muscle preparation
and holder are shown on a larger scale in Fig. 9.

which the preparation is prevented from drying by cover-
ing the bottom of the chamber with layers of filter paper
moistened with normal salt solution, while evaporation is
prevented by the glass shade. The nerve-muscle preparation and a convenient form of electrodes are shown on a
larger scale in Fig. 9. Below the moist chamber is seen

Fig. 9.

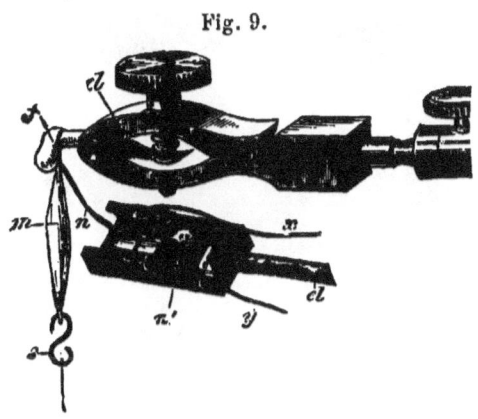

Enlarged representation of holder and electrodes seen in Fig. 8.
(From Foster's Physiology.)

a simple loaded lever by which the height of muscular
contraction, or the form of the muscle-curve, can be studied
by allowing it to record its movements on the smoked
paper of the revolving drum of the kymographion. Instead of the simple lever, which will of course, when elevated, describe an arc of a circle, Pflüger's myographion
lever (Fig. 10) may be employed. In very accurate
experiments non-polarizable electrodes should be used
(Fig. 11). These are readily made by plugging up one
end of a short piece of glass tubing a, with clay, b, moistened with normal salt solution; a few drops of saturated
solution of sulphate of zinc are then poured into the tube
as seen at c, care being taken that none flows over the
outside, and a strip of zinc, well amalgamated at the tip
and insulated by varnish over the remainder of its length,
is dipped into the zinc solution. The wires from the

battery are connected with the zinc; the electrodes may be conveniently held on pieces of heavy lead wire.

Fig. 10.

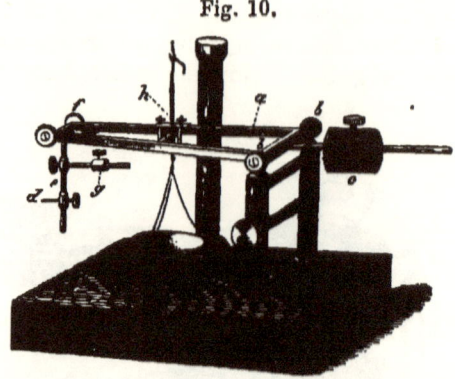

Pfluger's myographion. (From Sanderson's Handbook.) The lever *a* moves freely on the fulcra *b b*; at *f* is hung the swinging rod *e*, with the movable style *d*, and movable counterpoise *g*; at *c* is a heavy counterpoise to balance the lever. The tendon is connected with the lever by the thread *h*.

In many cases when the drug produces slight changes in the muscle curve, Marey's comparative myograph is

Fig. 11.

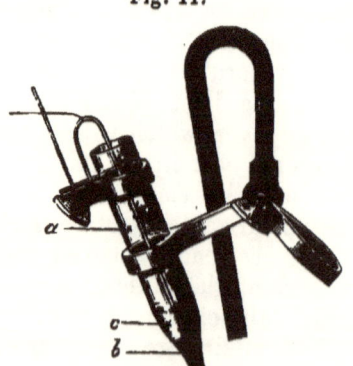

Non-polarizable electrode.

the most satisfactory instrument that can be employed; the mode of using it is represented in Fig. 12. A frog is fastened on its belly with pins, and the tendon of each gastrocnemius connected by a thread with a lever

Fig. 12.

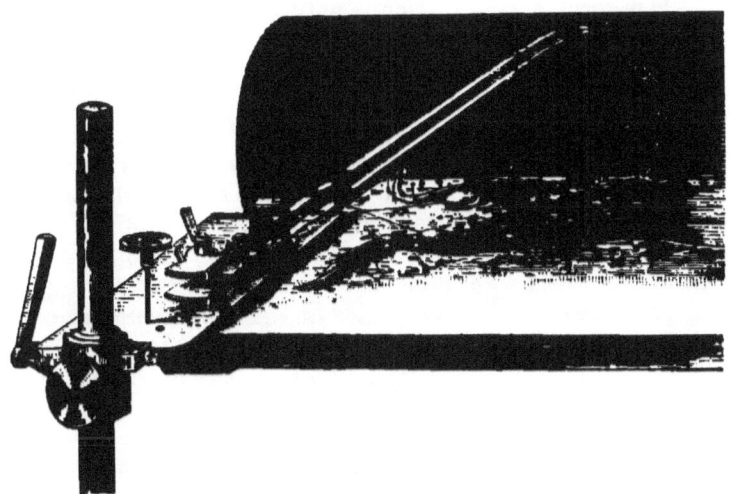

Marey's comparative myograph.

which writes on the revolving cylinder: the artery of one leg is then ligated and the poison injected into the body, and the muscle-curves, obtained by stimulating the poisoned and normal muscle, compared. The effects on the muscle in relation to indirect stimulation can also be studied by exposing the sciatic nerves and applying the electrodes to them instead of directly stimulating the muscle.]

In many cases where it is desired to determine simply the presence or absence of irritability, all that is necessary is to lay the nerve of the poisoned preparation on electrodes connected with an induction apparatus through which a strong current is passing. In many experiments the irritation may even be made by means of a small galvanic element consisting of a zinc and copper wire fastened together with a binding screw;[1] stimulation is

[1] Du Bois Reymond, Untersuchungen, 1 Taf. iii. fig. 19.

then produced by making or breaking the contact between the wires and the nerve.

2. EXAMINATION OF DIRECT MUSCULAR IRRITABILITY.
—This form of examination is only indicated when the indirect irritability is either entirely absent or greatly reduced. Muscles may be directly stimulated, either
mechanically, chemically (as by immersion in very dilute
acid), or electrically, as by bringing the electrodes of an
induction apparatus in contact with an isolated or exposed muscle. The rule in this case is to have the electrodes as widely separated as possible: therefore the
most convenient form of electrode consists of two flexible
insulated wires with their tips exposed, passed through
a few inches of flexible catheter and fastened together so
that about one inch of their free ends protrudes.

In all cases of direct irritation of a muscle, the nerve
fibres lying in the muscle are also subjected to the same
stimulation. If then direct stimulation of a muscle produces a contraction, even when indirect stimulation fails,
the contraction may be due to irritation of the intra-muscular nerve-fibres, since the nerve endings may preserve
their irritability even after paralysis of the nerve trunks.
By an experiment devised by Kühne[1] it may be determined whether this nerve irritability remains or not, it
being supposed, of course, that the muscle itself still preserves its irritability.

In the *sartorius* of the frog, the nerve enters the
muscle at its middle point, and sends branches towards
each end, but leaves a small portion at both extremities
of the muscle, near the tendon, entirely free from nerves.
If now a pair of electrodes, of which the tips are very
close together, are connected with an induction apparatus, and then placed perpendicularly on the *centre* of
a normal sartorius, it will be found that a much weaker
current will suffice to evoke contractions when the electrodes are at this point, than when they are over the por-

[1] Arch. f. Anat. u. Physiol., 1860, 477.

tions of the muscle near the tendon, which are free from
nerves. This indicates that the irritability of the nerves
is greater than that of muscle-fibre.[1]

If now, in a poisoned muscle, it is found that the difference in irritability at different points of the muscle is
absent, or is much less marked than in a normal muscle, it
may be concluded that the intra-muscular nerve-endings
have lost their excitability, or that it has been much
reduced.

3. EXAMINATION OF THE PHYSIOLOGICAL CONDUCTIV-
ITY OF MUSCLE.—When a normal muscular fibre is irritated at a single point, the resulting contraction travels
with great rapidity throughout its entire length. In
order to determine whether the muscle, when poisoned,
still preserves the property of transferring the contraction
throughout its entire length, a form of stimulation, such
as the mechanical, in which the stimulating effect is confined to a small, sharply defined area, must be selected ;
the experiment is also assisted when the simultaneous
irritation of the intra-muscular nerve-fibre is excluded.
The portion of the sartorius which is free from nerves is
therefore well adapted to the investigation of this point.
The poison is administered to a frog, and, when its effects
become evident, the sartorius is isolated and suspended
in an inverted position by its tendon of insertion, and the
lower end of the muscle, near its point of origin where it
is also free from nerves, divided transversely with a pair
of sharp scissors ; if the muscle is in its normal condition,
the mechanical stimulation of section with the scissors
will cause it to contract throughout its entire length.

Studies as to changes in the rate of progression of the
wave of contraction necessitate a complicated method, for
which the appropriate literature must be consulted.[2]

---

[1] Rosenthal, Moleschott's Untersuchungen, iii. 185.

[2] See Albey, Untersuchungen über die Fortpflanzungsgesch-
windigkeit, etc., Braunschweig, 1862. Bernstein, Untersuchungen über die Erregungsvorgang, etc. Heidelberg, 1871–78.

4. MEASUREMENT OF MUSCULAR ENERGY.—For this purpose comparative experiments must always be made on both poisoned and non-poisoned muscles, and it is advisable to employ excessively strong induction (tetanizing) currents, since both muscles must not only be subjected to the same stimulation, but must be stimulated simultaneously.

When it is not intended to measure the absolute energy, the muscle may be loaded with a medium-sized weight (about 20 grms. for the gastrocnemius), and the height to which the muscle, with a given stimulus, can raise the weight, determined ; then the weight is hung on the second muscle which is to be compared with the first, and the height of contraction, with the same stimulus, measured. The measurement of the height of contraction is most easily made by means of Pflüger's Myographion.[1]

An immediate comparison of the energy of two similar muscles may be made by hanging them alongside of one another and connecting their tendons with threads from which equal weights are suspended ; the threads are then wrapped once, in opposite directions, around a small pulley (a spool will do) furnished with a vertical index ; when both muscles are simultaneously stimulated with the same current, the index will move toward the weaker muscle.[2]

5. EXAMINATION OF ALTERATIONS IN THE CHEMICAL PROPERTIES OF MUSCLE.—Little more can be noted in this connection than the determination of the degree of rigidity of the muscle and of its reaction; both, of course, can only be taken into consideration when the muscle has completely lost its irritability. Rigor is easily recognized by the eye, especially when the thinner muscles are examined by transmitted light, since muscles in a condition of rigor become opaque; muscles in

[1] Pflüger, Electrotonus. Berlin, 1859, Taf. iii.
[2] Hermann, Diss. de Tono et Motor Muscul., Berlin, 1859; Nasse, Arch. f. d. ges. Phys., ii. 97.

a condition of rigor also lose their flexibility, and their rigidity could never be mistaken for convulsions. The most unmistakable sign of rigor, however, is the acid reaction developed, which can be readily recognized by touching a freshly cut surface of muscle with blue litmus paper. A general idea as to which of the albuminoids of muscles have, by their coagulation, produced the rigor, may be obtained by immersing the rigid muscle in 10 per cent. salt solution ; if produced by the coagulation of the spontaneously coagulable albuminoid myosin, the muscle will regain its normal flexibility, myosin being soluble in salt solution ; but if due to the coagulation of the albuminoids, which ordinarily are coagulated only by heat, acids, etc., no such change will be produced.

The color of the muscles should also be observed, as capable of throwing some light on the degree of impregnation with coloring matters and thereby giving some idea of changes occurring in the latter.[1]

For investigation as to this point, the blood must first be washed out of the bloodvessels of the muscles by injecting them with 0.5 per cent. salt solution. The qualitative examination of the muscle coloring matters (as to their percentage of oxygen, etc.), may be made by placing thin sections of muscle between two glass plates and then bringing them before the slit of the spectroscope.

For more precise study of changes in the chemical composition of muscle, no general rules can be given.

SECTION III.—**Action on the Nerves.**

With the exception of studies relating to the electromotor phenomena of nerves, to which the remarks on the study of the same points in muscles apply, isolated nerves need only be examined with reference to their power of transmitting motor impulses to the muscles in which they terminate.

[1] Kühne, Arch. f. Path. Anat., xxxiii. 79.

5

Possibly the examination as to the functions of secretory nerves, when excised with their glands, may be necessary to determine, by exclusion, what part of the results following their stimulation is not due to the presence or modification of their blood supply. All centripetal (sensory) nerves, on the other hand, can be studied only in connection with the central nervous system and its dependent motor apparatus, and their consideration must, therefore, be reserved for the chapters on the living animal, in which their modifications are also evidenced through changes in the character of reflexes.

At present, changes in excised motor nerves alone will be considered.

The examination of the irritability of motor nerves has already been alluded to in the study of indirect muscular irritability. In order to obtain any idea as to the state of irritability of a motor nerve, the muscle, with which it is in connection, must of course preserve its contractility unimpaired, or irritation of the nerve will be powerless to produce a contraction, and therefore no comparison of its irritability with a normal standard will be possible. If, however, the drug under examination produces paralysis of muscular fibres, the poison, while having free access to the nerve, must be prevented from acting on the muscle; this may be accomplished by ligating the femoral artery in one leg of a frog, near the knee (see p. 33), before the drug is administered, thus protecting the muscles of the leg from the poison. The sciatic nerve may then be tested as to its irritability with some prospect of reliable results. The readier method of immersing the nerve of an unpoisoned animal in the solution of the drug is unreliable, on account of the grounds mentioned on page 31, and results obtained by this method should be accepted with the greatest caution.

If it has been determined that the drug has no influence on direct muscular irritability, or if the irritability of the muscle has been maintained in the manner mentioned above, and it is then found that stimulation of the nerve fails to evoke a muscular contraction, the question then remains, what portion of the nerve, between the point of stimulation and its termination in the muscular fibre, has lost its function and thereby prevents the normal result of nerve stimulation? To appreciate the force of this it is only necessary to remember that every unirritable point in a nerve trunk is at the same time an obstacle to the conduction of nerve force; and in order to render a stimulation of a nerve of no effect, it is only necessary that any limited portion of the nerve below the point of application of the stimulus should have lost its irritability. To detect this point, therefore, successive points of the nerve are stimulated, gradually moving the electrodes down toward the muscle, until a contraction is produced, when it is known that the paralyzed point has been passed. If, however, stimulation of the nerve (with a weak current) at its point of entrance into the muscle fails to produce a contraction, it is evident, by exclusion, that the intra-muscular nerve filaments must be paralyzed. This supposition, again, may be verified by the mode of experimentation on the frog's sartorius, given on page 46.

Through the method given above only the lower limit of the region of paralysis can be determined; for the detection of the upper limit of paralysis of the nerve, no modification of this method of study, in which the contraction of the muscle is a measure of the irritability of the nerve, has yet been proposed.

The irritability of a nerve-fragment may be determined, without the use of its muscle, by the presence of the negative variation of the nerve current. When a normal nerve is irritated the natural nerve current suffers a negative variation on both sides of the point of stimulation, and through as great an extent of nerve as preserves its irritability. [This experiment is made by preparing as long a piece of sciatic nerve as possible and resting the cross section of the peripheral end on one non-polarizable electrode connected with a very sensitive galvanometer, while the other electrode is placed on its longi-

tudinal surface at some distance above ; another pair of
electrodes, connected with a tetanizing apparatus, is
then placed on the nerve as far as possible removed from
the galvanometer electrodes. If now, on passing a
current through the exciting electrodes, the deflection of
the galvanometer needle caused by the nerve current
does not undergo a negative variation, it may be known
that the paralyzed point lies somewhere between the
electrodes ; its exact limits may be determined by shift-
ing the electrodes.]

This method is therefore not only applicable here, but
also to the study of changes of irritability in sensory
nerves.

Section IV.—Action on the Heart.

[Of the other remaining organs which are capable of
isolated study, the heart is the most important and is
the organ which has received the greatest attention ; its
study in detail will, however, more appropriately be
taken up under the study of the circulation. At present
we will only indicate the methods of experiment which can
be performed as an introduction to the more extended
analysis of the action of the drug on the circulatory apparatus.

When kept in a moist atmosphere the irritability of
the excised frog's heart and its regulating nerves may be
preserved for hours ; the frog is therefore the animal
which is selected for these preliminary studies, though
the heart of the turtle, which also possesses these properties, may be used.

Changes in the rate of cardiac pulsation, the only
point which will be now considered, may be determined
in one of two ways : Either the drug may be applied to
the heart in situ, after destruction of the central nervous
system, or the heart may be excised and then brought
under the influence of the poison.

To perform the first of these experiments the frog is

"pithed" by cutting through the vertebral column immediately below the occipital bone and breaking up the central nervous system by passing a stout wire up and down
the vertebral canal; this operation should be so performed that little or no blood is lost. Should any
bleeding occur, it may be checked by driving a small
wooden peg into the cranium through the wound in the
vertebral column. The frog is then laid on its back and
the skin divided with scissors in the median line, and the
sternum and the upper anterior part of the visceral wall
removed so as to expose the pericardium, care being
taken not to injure the large abdominal vein which runs
in the median line on the anterior surface of the abdominal wall. The heart is then seen lying in the pericardial sac, and its rate of pulsation can be counted. A
snip is then made in the pericardium, and the pericardial
sac filled, by means of a pipette, with the solution of the
drug, and any changes in the cardiac rhythm noted.

For studies on the action of drugs on the rate of pulsation of the excised heart, two frogs of the same species
and of nearly the same size should be selected, and the
hearts exposed in the manner above described; the hearts
are then excised by cutting through all the cardiac vessels,
care being taken not to injure the sinus venosus, or any of
the chambers of the heart. One heart is then placed in
a watch-glass containing normal salt solution, and the
other, after its rate of beating has been determined, in
the solution of the poison; the changes in the rates of pulsation of the two hearts are then compared.] It should
be remembered that many gases, if the drug employed is
of such nature, and they are more suitable for this form
of experiment than are solutions, exert a more decided
action when locally applied than when acting through
the blood.

5*

PART II.

INVESTIGATION OF THE GENERAL ACTION
OF POISONS.

I.

SELECTION OF ANIMALS.

THE general rules given in this section have been drawn from the accumulated experience of previous experimentation, and though they may serve as a starting point for future investigations, they cannot be regarded as infallible. This reservation is especially applicable to indications given for the selection of animals as subjects for experimentation, since future investigations may show that many animals, now regarded as unsuitable for the experimental study of the action of poisons, may prove to be particularly valuable for the determination of special points.

The invertebrates, the warm- and cold-blooded vertebrates, or man, may serve as subjects for the investigations of the properties of drugs. Among these, the cold-blooded vertebrates are the most generally employed, and the same peculiarities which have made the frog a martyr to physiology make him subject to martyrdom in the cause of pharmacology. For not only may the frog withstand the most severe operations, but also nearly all its organs may be removed from the circulation and still preserve their functions unaltered.

In the warm-blooded animals, all poisons which, through action on the heart, interfere with the circulation, or which, through interference with respiration, alter the proportion of gases in the blood, at the same time produce extensive alteration in the functions of all organs ; while the same drugs administered to frogs may fail to produce any general disturbance, particularly in the nervous system or muscles. When, therefore, after the administration of any poison, departures from the normal functions are observed in the muscular or nervous system of the frog, they may be ascribed with certainty to the direct action of the poison on these organs, without the possibility of their being attributable to some secondary reaction following from disturbance in the circulatory or respiratory apparatus, or in changes in the respiratory functions of the blood. In the closest connection with this peculiarity of the frog, may be mentioned, as has been already alluded to in the preceding pages, the possibility of excising or excluding from the circulation (the latter of the greatest importance, where it is desired to protect certain tissues from the action of the poison) the various organs of the frog without their suffering any profound disturbance in function. If any further advantages were needed, it might be added that in frogs the heart is extremely easy to expose, and that the circulation is readily accessible for microscopic study.

Nevertheless, the investigation of the action of all poisons on warm-blooded animals is in the highest degree important. The essential object of pharmacology is in all cases a recognition of the action of poisons on man, and it cannot be denied that experiments on warm-blooded animals furnish means of deducing more reliable conclusions as to this point than when made on frogs. In addition the following reasons render advisable, on physiological grounds, the study of poisons on mammals. In the first place, the small size of many organs in the frog, such as the different glands, renders an accurate study impossible ; while in the larger mammals the ducts of most of the glands are readily isolated, and the secretions measured and obtained in sufficient amount for chemical examination. Then the location of deposit of poisons in the system, their excretion, and the chemical changes which they undergo, can, on account of the quantities of material necessary for chemical study, only be determined in the larger animals. While, therefore, on these grounds, warm-blooded animals as a class are required for the thorough study of the *modus operandi* of poisons, as a rule it makes no difference which members of this class, with certain special exceptions to be hereafter mentioned, are chosen. Before all things, however, it must be emphasized that the general action of all poisons which act either on the blood, the heart, or the respiration, may differ in the most marked degree, even when the action on individual organs closely corresponds ; this is readily explainable on the grounds already mentioned, and the points of difference will be subsequently more closely studied.

Then again, there are many poisons which have a different action on different groups of animals, and to obtain a complete idea of the action of such poisons, various types of animals must of course be experimented on.

These are then the chief points which render the warm-blooded animals indispensable for pharmacological studies. From what has been said it follows that the last of the above-mentioned points may be worked out on the smaller animals, such as small birds, pigeons, mice, rats, guinea-pigs, etc., while the larger mammals, such as cats, rabbits, dogs, or horses, besides being more easily operated on, are best suited for the first-mentioned purposes.

For example, suppose that we are dealing with a poison which has been found to paralyze the muscles of respiration in frogs, and in mammals, in addition to the respiratory paralysis, to cause convulsions ; it must then be determined whether the convulsions are only secondary to the interference with respiration ; this would explain their absence in the case of the frog. The simple experiment of artificial respiration will decide this in the case of the mammal, since the removal or prevention of the respiratory disturbance, in such instances, will remove or suspend the cause of the convulsions.

Then too, measurements of blood pressure and the intravenous injection of poisons, can be readily practised only on the larger mammals.

On these grounds, therefore, rabbits and dogs are the examples of warm-blooded animals ordinarily used for pharmacological purposes. They furnish representatives of the two groups of the herbivora and carnivora, on which many drugs act in different degrees of intensity. Cats, on account of their cheapness, would be largely preferred to dogs, if experimenting on them was less unpleasant. Horses, unfortunately, can be but rarely obtained for pharmacological work; hens, as large and cheap respresentatives of birds, are very valuable for the study of many points, though it must be remembered that many poisons, e. g., strychnia, act in different degrees on birds and mammals, on account of the chemical diversities in the renal secretions of these animals. The smaller animals, such as mice, rats, guinea-pigs, and pigeons, may be used to determine the general action of the poison on warm-blooded animals. Other animals may on special grounds be valuable subjects for study; such as the bat, on account of its pulsating wing-vessels, rabbits on account of the readiness with which the circulation, especially after depilation, may be studied in the vessels of the ear, and albino-rabbits on account of the peculiarity of their eyes.

Associated with experimentation on warm-blooded animals, the determination of the action of most poisons on man is of the highest interest, since in man only can the group of phenomena produced by action on the sensory apparatus be studied with any prospect of obtaining reliable results; any correct idea as to toxicological doses, data of the greatest practical importance, moreover, cannot, as a rule, be obtained from experiments on animals. All modifications of sensibility and consciousness, and possibility of voluntary motion, can be examined in experiments on ourselves, and the results recorded either during the experiment or from memory afterwards, or experiments can be made on another and the results communicated during the course of the experiment. At the best, however, the comparison between the drugged and normal condition, though more reliable than similar modes of investigation on animals, is apt to be very deceptive.

The invertebrate animals have as yet been very seldom employed in pharmacological studies ; general rules as to the condition in which they are suitable cannot, therefore, be given, it being, however, understood that since these animals only possess hæmoglobin in exceptionally small quantities, they cannot be used for the study of drugs which act on that tissue.

From the above it may be concluded that methodical investigations of the action of a poison should commence with experiments on frogs, and then on warm-blooded animals, and later, in suitable cases, on man, if the cases of poisoning which have been recorded in the case of nearly all poisons do not sufficiently clear up those points which are alone suitable for study on man.

[The important position recently assumed by the lower organisms, such as bacteria and bacilli, in disease processes, renders it advisable to test the action of all new drugs, and drugs imperfectly understood, on these organisms ; but since the whole subject is at present in a transitory condition, little more than an outline of a few rudimentary experiments can be given.

The action on human protoplasm may be assumed to have been already studied with the white blood-corpuscles ; protoplasm in lower forms may be studied in the action of drugs on the amœba, often to be found in moist, damp earth; or on infusoria, found in great variety in all fresh-water infusions of decaying animal or vegetable matter. The limit of study drawn by our present lack of knowledge lies in determining whether the movements of the infusoria, bacteria, etc., are checked by the drug, or whether the drug furnishes a good medium for their development; or, perhaps, it may be found that certain substances added to the culture fluids in which the specific bacilli or micrococci are being developed, may destroy their virulence without destroying their life, or may entirely prevent their appearance.

Infusoria may be obtained in great variety and numbers by making an infusion of hay, or other vegetable matter, and allowing it to stand for several days in the open air. A drop of the fluid is then withdrawn with a pipette and placed on a microscopic slide, covered with a slip, and examined; after infusoria in active motion are found, a drop of the solution is run in under the edge of the cover-slip, and the field again examined to see if the infusoria are still moving. If they appear unchanged, a drop of a stronger solution is passed in, and the field again examined, and so on, until the necessary strength of solution is determined, or until it is proved that the drug is without effect. If the substance experimented with should be a gas, a drop of the infusion should be placed on a cover-slip, and the latter placed on Stricker's gas chamber, which is to be used in the ordinary manner.

As discovered by Binz, certain substances, such as various salts, may destroy the life of infusoria by the abstraction of water which they cause, while others, particularly such substances as are known as antiseptics, destroy low forms of life in some unknown manner.

Bacteria and vibrios may be readily obtained in infusions of putrefying animal matter, such as boiled white of egg, or meat, and the influence of drugs upon them may be tested in the same manner as on infusoria.]

Modes of Securing Animals.

[Nearly all the methods of investigation of the action of drugs hereafter to be described require that the animals, on which the experiments are made, should be prevented from moving. This restraint may be of two kinds, either mechanical or chemical. The latter method, that of narcotization, is less often requisite; the conditions under which it is permissible will be given in the chapter on anæsthetics.

Frogs are fastened to a board, about nine inches long by three broad, by slip-knots passed over each elbow and ankle, while the ends of the cords are secured at the corners of the board, either by wrapping around cleats, or by passing them through binding screws such as are ordinarily used in electric batteries. Or holes may be made through the corners of the board, and the ends of the cords passed through them, and prevented from slipping either by plugging the holes with small wooden pegs, or simply by tying the ends of the cords together.

Rabbits and cats are best secured with Czermak's rabbit-holder (Fig. 13). The bar h is first passed be-

Fig. 13.

Czermak's rabbit-holder.

hind the animal's incisor teeth (the canines in the case of cats), the bar g being below the lower jaw, and the screw k then tightened until the jaws are so tightly compressed that the bar h cannot slip beyond the incisors. The limbs are then fastened by the slip-knots at m, and the end of the holder pushed into the fork at f, and screwed fast. The rabbit may, of course, be fastened either on its back or belly, as the conditions of the experiment require. Guinea-pigs may be secured with the same apparatus in the same manner, it being, however, generally necessary to pad the bars g and g' with

6

Fig. 14.

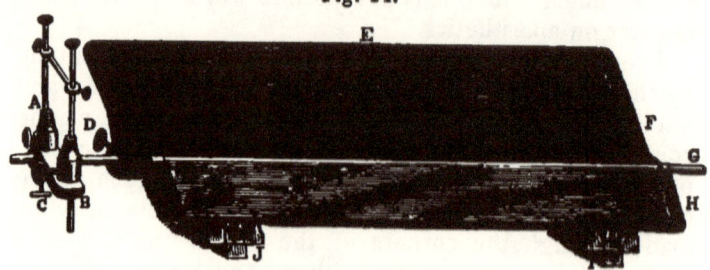

Bernard's dog-holder.

Fig. 15.

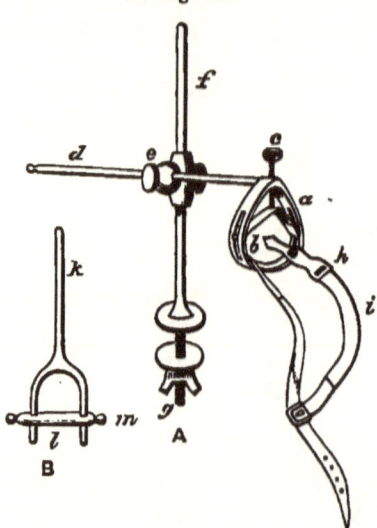

A. Improved form of Bernard's dog-holder. At *h* is a straight piece of metal which passes beneath the dog's lower jaw and is connected with the strap *i*, which is fastened behind the head. The holder may be moved backwards or forwards by sliding the rod *d* through the clamp *e*, or up and down by moving *e* on the iron rod *f*, which is fastened to the table by the nut *g*.

B. Brunton's holder for dogs or rabbits. A loop of cord is tied around the upper jaw, the bar *l* passed behind the canine teeth of the dog or cat, or incisors of the rabbit, and the two jaws tied together to prevent its slipping out. The fork *k* is then pushed through the holes in *l*, and fastened by the screw *m* ; *k* may then be fastened to an upright bar as in Czermak's or Bernard's holders.

cotton, so as to prevent them from slipping over the head.

Dogs may be secured with Bernard's holder (Fig. 14), the bar *e* being passed behind the canine teeth, and the jaws fastened together by means of a cord tied around the muzzle, or the apparatus represented in Fig. 15 may be used. The muzzle is passed through the ring, and the strap *h, i* is tightly buckled behind the head, and the screw *c* tightened up, so as to prevent the jaws being opened, while the strap prevents the head being retracted ; the animal is fastened in any desired position by slip-knots passed around the legs and tied into the different holes in the side-boards. It is often advisable, when very large dogs are used, to pass the cords with which the forelegs are tied under the animal's body, and then make the ends fast on the opposite sides of the box ; by this means the forelegs are held tightly against the animal's sides, and the dog is rendered immovable. For methods of fastening the larger mammals and birds, reference must be made to Bernard's *Physiologie Opératoire*, or to Livon's *Manuel de Vivisection.*]

II.

ADMINISTRATION OF POISONS.

IF it is desired to study only the local effects of a poison, no special directions are required, it being simply necessary to observe the general rules that the solvent employed must itself have no irritant action on the part to which the poison is applied, and that the application should be several times repeated if the first administration appear to be without effect. As instances of the effects which might be obtained by neglect of these precautions, we might mention the fallacy of deducing results from the application of an alcoholic solution to a mucous surface, the injection of a fluid into the lungs to determine the local action on the bronchial mucous membrane, or finally from the single application of tartar emetic to the human skin. In nearly all cases control experiments with the solvent alone should be made, so as to exclude the possibility of any of the results observed being attributable to it.

At present we will consider only the methods of administering drugs for the determination of their general action.

The general action of a poison, as will be found later, commences as soon as the substance enters the circulation, and lasts as long as it remains in the blood in an unaltered condition. For the production, therefore, of its general action, the poison must be introduced in some way or other into the blood, either in the normal manner by absorption through the closed walls of the bloodvessels, or by being directly introduced into the circulation through an opening in the walls of a vessel.

Physiological absorption always occurs gradually and with varying degrees of rapidity, depending upon the locality of the absorbing surface and the nature and solubility of the poison; while by direct injection of the substance into the bloodvessels any amount desired may be allowed suddenly to produce its characteristic action.

Absorption occurs most slowly when the drug is brought into contact only with the external skin; more rapidly with mucous surfaces and the internal surfaces of serous sacs, such as the lymph-sacs beneath the skin of the frog; and with still greater rapidity when brought in contact with wounds in the skin or mucous surfaces, or when introduced into the subcutaneous connective tissue (hypodermic injection). These last three methods have the point in common that by the anatomical lesion the substance is brought into direct contact with the walls of the capillaries, and possibly with openings in their walls, while in the other methods the substance is separated from the bloodvessels by an epithelial layer which is only traversed with comparative difficulty. By the selection of one of these methods, therefore, it is possible to regulate the rate of action of the poison, and obtain any varying degree of rapidity from the instantaneous action of substances injected directly into the vessels, to the extremely slow and gradual action when the poison is brought in contact with the external skin.   It must, however, be emphasized that not only the rapidity of appearance of the general action but also the existence of *any* general action may depend upon the mode of administration.   It will be shown later, that the general action of every poison depends upon the presence of a certain amount of that substance in the blood, and that the amount circulating in the blood depends not only on the amount administered but upon the rapidity with which excretion takes place ; and it can accordingly happen that a poison which in itself is very active may from slow absorption and rapid excretion not attain a sufficient amount at any one time in the blood to produce any effect.

Since, then, by the choice of the mode of administration we can accelerate, hinder, or prevent the production of the general action of a poison, and since the rapidity with which the symptoms appear has a marked influence on the general picture of the poisoning, and since many poisons may be chemically altered by the peculiarities of the absorbing surfaces, it is advisable to perform a number of experiments in which these different methods of administration are employed.

SECTION I.—Injection into the Bloodvessels.

In the case of many poisons the action will differ, depending upon whether it is injected into an artery or a vein.   As a general rule it is preferable to inject the drug into a vein ; in the first place, on account of the readiness with which large veins, from their superficial location, can be isolated ; second, on account of the sensibility of arteries ; and third, and most important of all, because substances injected into a vein are first carried to the heart and then gradually carried to all parts of the system. Injections into the venous system are most easily made in mammals into the external jugular vein, and preferably towards the heart, since in this direction larger quantities of fluid can be thrown into the circulation than when it is injected towards the periphery where the resistance of the valves and of the capillaries may easily prevent the entrance of fluid. In the frog, only the large venous trunks are suitable for injections.

In all injections, especially when made into the central end of the jugular vein, the entrance of air must be avoided with the greatest care.

The external jugular vein is exposed in the following manner: After cutting the hair from the parts, a short incision is made through the skin of the neck in a line drawn from the angle of the jaw to the jugular notch in the sternum; then, on carefully clearing away the fat and connective tissue with two pairs of forceps, or with a blunt hook, the dark blue vein, usually sending off numerous branches, is exposed to view. The vein is then to be carefully freed from its connective-tissue sheath for a length of about 2 cm., and a double thread passed beneath it with a curved, blunt needle; the central end of the isolated portion is then to be compressed with a spring-clip, or with a loop tied in a single bow-knot, and the peripheral end ligated with one of the threads previously passed around the vein. The portion of the vein lying between the ligature and the clip (or loop), which, from the order of applying the ligatures, is distended with blood, is then snipped with a pair of scissors cutting well at the points, and a glass or metal canula inserted and tied in by the remaining thread.

The canula can be conveniently furnished with a stopcock, and, before insertion, must be completely filled with distilled water, which is prevented from flowing out by closing the cock. After the insertion of the canula, the spring-clip on the vein can be removed. The poison is to be injected by means of a metal or glass syringe, conveniently graduated into cubic centimeters, whose nozzle must accurately fit the mouth of the canula. [If a glass canula is used, it is only necessary that the free end be inserted into a piece of rubber tube, about an inch long, which fits the end of the syringe. It is then not necessary to fill the canula until it is desired to make the injection, when it is advisable to fill the canula *completely* with the solution to be injected, instead of water ; of course, then, the clip must not be removed from the vein until after the syringe is bound into the rubber end of the canula.]

The injection must be made slowly and gradually, and care must be taken, by holding the canula, that the syringe does not separate from the canula by the force necessary in making the injection ; it is, therefore, an advantage to use a syringe on the upper end of which are two rings for the insertion of the second and third fingers, while the thumb is passed into a ring in the end of the piston-rod, so as to admit of one hand manipulating the syringe, while the other holds the canula. When it is desirable to inject the poison into the arterial system, the carotid or femoral artery is selected, unless there is some special reason for the use of other vessels. The carotid artery is exposed by an incision parallel to the trachea, passing between the sterno-hyoid and sterno-mastoid muscles; it lies in the common sheath with the vagus, sympathetic, etc. The crural artery is readily found in the inner aspect of the thigh, below Poupart's ligament, where it lies very superficially and may be felt through the skin.

Canulæ similar to those employed in the veins may also be used for arterial injections, either toward the heart or the periphery ; in both cases considerable resistance is met with, from the general arterial tension in the first case, and from the capillary resistance in the second. Injections towards the periphery invariably evoke the symptoms due to the poison first in the organs supplied by the artery, while when injected toward the heart, this is the organ first affected, and then the entire system is gradually brought under its influence. This latter method is to be preferred unless some local effect is to be studied, though it should be remembered that a marked accelerative effect on the heart can be thereby produced. Canulæ are inserted into arteries in precisely the same manner as into the veins.

Injections into arteries have the disadvantage that in most cases they are extremely painful, and that the evidences of pain on the part of the animal may be confounded with, or mask, the specific effects of the drug. In such cases, therefore, if possible, a control experiment must be made after the narcotization of the animal.

It cannot be too much emphasized that by no means all poisons are suited for direct injection into the circulation. It is evident that solid particles, even if suspended in a fluid, cannot be injected into the blood, since they would remain as emboli at the narrower subdivisions of the vessels. Nor, again, can fluids be injected which produce emboli by causing a precipitate in the blood. It would, for example, be an extremely defective experiment to inject a mineral acid into the blood, since the coagula thereby produced would be swept into the circulation, form emboli, and produce symptoms which would in no way depend upon the specific action of the substance injected.

It must, therefore, not be forgotten that the object of the injection of a poison into the blood is not to determine the action of the poison on the blood, but only to evoke, in the most rapid manner, the general symptoms due to the poison, after having first excluded by experiment every possible alteration of the drug before absorption occurs; therefore, an injection should never be made into the circulation until it is known what changes, if any, occur either in the blood or drug when the two are brought in contact (see p. 20).

On the other hand, even when it has been found that the poison causes a precipitate when mixed with the blood, it may be necessary, in exceptional cases, to inject that poison into the blood to settle the question as to

whether the general toxic effects of the substance otherwise administered are due to the production of emboli.

Aëriform poisons are also unsuitable for intravascular
injections, although pharmacological literature is rich in
such experiments. In the first place, the entrance of
any gas, even should it be considered most absorbable,
into a bloodvessel, by producing gaseous emboli, can
evoke the same symptoms as follow the entrance of air
into the veins. In the second place, there can be nothing
gained by injecting a gas into a bloodvessel, even if
this danger could be avoided ; since a gas cannot possibly
enter the blood more rapidly than by absorption through
the lungs, where a much larger surface of blood is exposed to its influence than when it is injected into a vessel. For example, it is not possible to inject oxygen
into a vein in sufficient quantity and rapidity to prevent
the appearance of dyspnœa, if other sources of oxygen
are withheld. Here, however, as in the first instance of
unsuitable injections, exceptional circumstances may render the administration of gases through the lungs impossible, as when it is desired to determine whether the
poison is eliminated through the lungs ; even in such a
case, however, the subcutaneous injection of the gas is
preferable to its direct introduction into a vein.

If the gas is readily soluble in water, there is no
objection to the injection of an aqueous solution into a
bloodvessel, or a quantity of blood may be drawn from
the animal, defibrinated, treated with the gas, and again
injected after being warmed up to the body temperature.
For such an experiment it may be advantageous to make
use of a T-formed canula, so that the blood may be drawn
from one end of the vessel and injected into the other.
A simple canula, fastened into the central end of a vein,
will, however, answer every purpose, as sufficient blood
can be drawn from the vessel by a syringe inserted into
the canula.

SECTION II.—Subcutaneous Injections and Injections into Serous Sacs, Lymph Sacs of the Frog, etc.

Graduated syringes, such as employed in injections into veins, may be used, substituting for the canulæ then used, the ordinary hypodermic needles, or tubes with sharpened points ; after the injection has been made by pinching up a fold of skin and inserting the needle nearly horizontally, the needle and syringe are withdrawn together, and the wound of entrance of the needle closed for a few seconds with the finger to prevent the escape of the fluid ; it is also advantageous to draw the skin, before the injection, to one side, so that, the wounds in the skin and connective tissue not coinciding when the skin regains its natural position, a valvular opening is made and the escape of fluid further prevented.

[In frogs the solution may be injected into the abdominal cavity or under the skin of the back, when it will find entrance into the large dorsal lymph-sac, and thence be rapidly carried into the general circulation.   In rabbits, guinea-pigs, and dogs, it may be injected under the skin of the flank, though it is only convenience in handling which renders this locality to be preferred to any other ; in guinea-pigs the abdominal walls are extremely thin, so, if this situation be used, care must be taken, when comparative experiments are made, that the abdominal cavity is not penetrated by the needle, when, of course, absorption would be much more rapid.]

SECTION III.—Insertion of Poisons into the Mucous Cavities of the Body.

In many cases, poisons may be introduced into the stomach of animals by mixing them with the food ; usually, however, it is better to inject them, especially when fluid, by means of a stomach tube.   In rabbits, an ordinary medium-sized flexible catheter will answer this purpose; for its introduction the animal is held with the head well extended and the jaws tightly closed, so that it cannot bite the catheter, which, after being well oiled, is passed into the mouth behind the incisor teeth and pushed well backward, when ordinarily it will pass without difficulty into the œsophagus. If it should pass into the trachea, as would be rendered evident by the resistance of the vocal cords and the occurrence of dyspnœa on closing the catheter, it must be withdrawn and reinserted. The end of the catheter must be furnished with a short piece of rubber tubing which will fit the nozzle or canula of the syringe, and the syringe and catheter must be withdrawn together after making the injection [otherwise, the catheter remaining alone would act like a siphon and allow the contents of the stomach to escape]. Œsophageal sounds, similar to those employed in man, can readily be introduced in dogs over a block of wood so inserted as to keep their jaws wide apart:—this block may conveniently have a hole bored in its centre for the passage of the sound, and two straps at its ends to buckle behind the head and so keep it in position. A similar operation is readily performed on cats, making use of a smaller wooden or iron bar to keep the jaws open.

In frogs, after opening the mouth with the handle of a scalpel, an ordinary canula can be passed into the stomach, and the injection made through it. Solid bodies, such as crystals, can be readily pushed into the stomach by means of a glass rod.

Applications to the rectum, bladder, etc., are performed in the ordinary well-known manner. Instillations into the conjunctival sac are best made by drawing the lower eyelid away from the eyeball with the hand or a pair of forceps, and then dropping the solution by means of a pipette, or medicine-dropper, into the open sac; it is well to maintain this everted position of the lower eyelid for some little time, so as to permit of absorption before much of the fluid is lost by closure of the eyelids. This method is only suitable for studying the local action of the poison.

The bronchial mucous membrane is almost never to
be used for the absorption of solutions.   Should it,
however, be necessary to make the experiment, a canula
can be readily introduced into the windpipe by performing tracheotomy, and the poison injected.   A great deal
of force will be required to inject the solution into the
bronchioles.

SECTION IV.—Administration of Gases and Vapors through the Lungs.

This method of administration of poisons is very often
desirable, and can be accomplished in several different
ways, depending upon whether the gas may be mingled
with the inspired air or must be separated from it.

Small animals, such as frogs, birds, and the smaller
mammals, may be placed in a chamber filled with the
gas either before or after the introduction of the animal.
In the first case, when frogs are employed, it is only
necessary to float a bell-jar containing the gas on water
or mercury, and then pass the animal through the water
into the holder; this method may also be employed for
other animals, but it has the disadvantage of chilling the
animal from the necessary wetting; on the other hand, it
has the advantage of plunging the animal suddenly into an
atmosphere of the gas whose effects it is desired to study;
the air contained in the animal's lungs, particularly in
the case of frogs, may be largely expelled by pressure
under the water.

The other method consists in placing the animal under
a bell-jar, or in a cylinder, through the top of which the
gas is passed by means of a tube reaching to the bottom
of the vessel, while the air is allowed to escape through
a second opening in the top.   In order to regulate the
entrance of the gas, the bottom of the receiver may be
covered with a layer of water, under which the tube of
entrance dips, so that the rate of entrance may be known
by the rapidity with which the gas-bubbles pass, while

the animal may be placed on a board which either floats on the surface of the water or is raised above it by supports. This method is, however, only applicable when complete exclusion of atmospheric air is not essential.

In most cases it is much more advantageous to connect the animal's lungs directly with the gas apparatus. To accomplish this, it is not always necessary to perform tracheotomy, as the head may be covered with a rubber bag or cap which is in communication with the gas receiver. In rabbits, when it is desired to avoid tracheotomy, the head can be passed into a funnel, fastened by appropriate bands, while the space between the neck and the funnel can be plugged almost air tight with cotton or wool.

For experiments on man, it is only necessary to insert a tube, through which the gas is conducted, into the mouth, while the nostrils are closed.

As a rule, it is preferable to perform tracheotomy on animals, since then the exclusion of air is assured and influence on the glottis is avoided, thus rendering experiments with irrespirable gases possible. Tracheal canulæ, of the same shape as the venous canulæ, but of larger size, can be employed, though T-shaped glass or brass tubes are better.

Modifications of this method will be necessary if it is desired to separate the air of inspiration from that of expiration. In any case, however, it is only permissible to allow the animal continuously to inspire and expire into a closed gas-chamber, when the vessel is so large or the duration of the experiment so short, that the increasing contamination of the gas from the products of expiration and absence of oxygen is not to be feared. If the gas-holder is so large in comparison to the thorax of the animal that the variations of pressure on inspiration and expiration are very small, and it is not necessary that the volume of the holder should change with the phases of respiration, a simple gasometer or even a large bell-jar floating in water with an opening connecting with the tracheal tube will answer. As a rule, however, the gas-holder should readily decrease in volume in inspiration

7

and increase in expiration; such gasometers may be
represented by rubber or silk bags, or by a Hutchinson's
spirometer when the air-chamber is accurately balanced
and care taken to expel all the air from its interior.

In the most careful experiments the inspired air should
be separated from that expired. For this purpose the
tracheal canula, or the tube of the funnel used as a mask,
must be connected with a fork-shaped tube, of which one
prong must be furnished with a valve opening only
towards the interior, the other with a valve opening only
towards the exterior. For this purpose, there is nothing
better than Müller's liquid valves which are perfectly
efficacious in forming tight joints and which can be
readily prepared.[1] These consist of bottles (with one
or two necks) through the corks of which two tubes are
passed, one passing merely through the cork, the other
to the bottom of the flask, and dipping below a layer of
liquid, either water or mercury. Gases can then only be
passed through the flasks from the longer to the shorter
tube, provided that the pressure is not sufficient to raise
a column of the liquid used higher than the length of the
longer tube. Two of these flasks are then so arranged
that the inspiration flask has its shorter tube, and the
expiration flask has its longer tube, connected with the
forked-shape tube of the tracheal canula: the length of
the long tube must be proportionate to the strength of
respiration in the animal, and when mercury is used as a
valve it can naturally be much shorter than when water
is used for the same purpose; hence when dogs are
used and where quantitative analyses of the expired gases
are made, particularly of such gases as carbonic acid,
which is soluble in water, it is always advisable to employ mercury.

The long tube of the inspiratory valve is then connected with the gas-holder, for which either a spirometer, or Rosenthal's modification of Bunsen's gasometer,
when it is desired to shut off the gas-supply through mer-

---

[1] Annalen der Chemie und Pharm., cviii. 257.

cury, may be used.[1] The short arm of the expiratory valve either opens into the air or into a gas-holder when it is desired to examine the expired air.

If it is desired to mix small quantities of the poisonous gas with the atmosphere before inspiration, the long tube of the inspiratory valve may communicate with the atmosphere, and a second long tube in the same flask may be connected with a gas-holder or bag.[2]

If experiments are being made with a gas which is freely soluble in water, or with the vapor of volatile liquids, the solution of gas in water or the volatile liquid may replace the fluid which acts as a valve in the inspiratory flask, which, if necessary, may be warmed to facilitate its vaporization. So, also, the fluid in the expiratory flask may be replaced by, or covered by, a layer of reagents, which, by their chemical behavior, will show the presence of certain substances in the expired air.

In all experiments in which respiration is carried on by the animal, the tubes in these valves should be wide, and not too long, so as not to cause dyspnœa from resistance to respiration. In many cases, it is preferable to force the gases into the animal's lungs, rather than to depend upon its own respiratory powers, thus rendering it possible to open the thorax and inspect the heart, and to study the action of the gas after the production of apnœa. The principal difficulty met with in accomplishing this lies in the production of expiration. Hook's method was to neglect expiration entirely, and to pass the gas continually through the lungs by opening the thorax, and then incising the lungs in several places; apart from the loss of blood entailed by this proceeding, it has the further disadvantage, that, even when numerous openings are made in the lungs, the entire organ is not permeated by the gas, since many portions collapse.

[1] Rosenthal, Arch. f. Anat. u. Phys., 1864, 456.
[2] Kaufmann u. Rosenthal, ditto, 1865, 659.

Passive or active expiration can be produced in uninjured lungs in one of two ways: in the first, the tracheal canula has an opening at one side, through which
the lungs, by contracting from their own elasticity, are
enabled to force out the expired air during the pauses
between inspirations; this opening, by graduation of its
size, has the further function of acting as a safety valve,
and preventing rupture of the lungs by too great a pressure in inspiration. This opening can also be, with advantage, connected with a Müller's valve, and the resistance can then be regulated at pleasure. When extremely
poisonous gases are being experimented with, this opening, or the short tube of the valve, should be connected
with a tube by which the expired air may be conducted
out of the laboratory.

The other method consists in interposing between the
lungs and the gas-apparatus a two-way cock which works,
by some mechanism, in the rhythm of respiration. For
this purpose the fork-shaped tube may be connected
with two rubber tubes which are alternately compressed
by a lever.[1]

For the production of artificial respiration, any of the
various forms of blast apparatus, described in all hand-
books for the laboratory, may be made use of. When a
constant force only is required, the gas may be passed
directly from the generator; but where a rhythm is
required for forcing atmospheric air, an ordinary bellows
worked by hand, gas, or water-power, will answer every
purpose.[2] When bellows are used to force gas other
than air into the lungs, the entrance valve of the bellows
must be connected with the gas-holder, while the nozzle
of the bellows communicates with the lungs. Various other
arrangements for producing a constant or rhythmical
air blast will readily suggest themselves if the above are
not accessible; such as an ordinary rubber bulb arranged

---

[1] Rosenthal, Arch. f. Anat. u. Phys., 1864, 456.
[2] Thiry describes a simple form of automatic respiration apparatus in Recueil des Travaux de la Soc. Méd. Allemande, Paris,
1855, 55.

with valves acting in the same direction, and compressed rhythmically under the foot.

One other arrangement is necessary to complete the apparatus for the study of poisons administered through the lungs. In order to obtain an uncomplicated result from the action of any poisonous gas or vapor, it is necessary that it should be suddenly substituted for the atmospheric air; if the animal is breathing spontaneously, it suffices to suddenly connect the tracheal canula with the inspiratory valves connected with the gas-holder, as can be readily done with a three-way cock, one opening of which communicates with the atmosphere. In artifical respiration the three-way cock can be inserted between the blast and the gas-holder, an arrangement which will permit the introduction of the gas after the production of apnœa by forced inspiration.

Other modes of application do not require any special mention.

III.

INVESTIGATION OF THE PATHS OF ELIMINATION AND CHANGES OF POISONS IN THE BODY.

From the great diversity of poisons, only a few general rules can be given for the elucidation of these points; as a rule, the most various chemical manipulations are requisite. It has been already said that the general action of a poison can only appear after its entrance into the blood, and since many poisons, for which there are no known chemical tests, produce marked symptoms in extremely minute quantities, the appearance of the characteristic effects is in most cases a far more delicate test of the presence of the poison in the blood than can be obtained by any chemical reagent. Nevertheless the appearance of certain symptoms after the administration of a poison does not exclude the possibility of some change in the constitution of that poison after absorption, as the products of decomposition of the poison may themselves have been the cause of the symptoms ; therefore it is advisable to attempt the recognition by chemical means, difficult though it be, of the presence of the drug in the blood. In all cases, particularly where organic poisons are concerned, the blood should be examined as soon as possible after withdrawal from the body ; in many cases when the recognition of the poison in the blood fails, the poison can be detected unaltered in the secretions, such as the saliva or urine.

The greatest diversity may exist as to the ultimate fate of a poisonous substance introduced into the blood ; and this diversity will be more marked the more rapidly death puts an end to the tissue changes.

SECTION I.—Passage of Drugs through the System without Change.

The usual fate of foreign substances introduced into the organism is that they sooner or later, in different manners, are excreted from the body, while their extended or permanent deposit in the tissues is exceptional.

There are two organs which especially have for their function the removal of deleterious substances from the blood ; the kidneys for fluids or solutions, the lungs for gases and vapors. In the former, the blood loses large quantities of fluid, which, from the laws of diffusion, must carry with it appreciable quantities of the substances in solution in the serum. In the latter, at least during respiration, large blood surfaces are in contact with the atmosphere, and since this atmosphere at best can contain but traces of the deleterious gaseous substances in the blood, and therefore under less tension than in the blood, they tend to constantly diffuse, under ordinary physical laws, from the blood into the air in the lungs. Exceptional circumstances may, however, occur in which these physical processes will be either hindered or facilitated. For example, diffusion will be prevented by chemical union of the foreign bodies with the ingredients of the blood, or accelerated by chemical forces tending to displace them from the blood. As an illustration of the former condition, carbon monoxide may be mentioned, which, from its chemical relations with the blood, shows no tendency to diffuse into the air in the lungs ; while as an example of the second, carbon dioxide, though forming a chemical compound with the blood, is diffused with the greatest readiness. Similar conditions exist as regards the elimination of substances through the kidney.

As in the kidneys, so also in all the other emunctory organs, there is, probably, a tendency to the elimination of foreign substances from the blood ; therefore drugs, after absorption, may appear in the saliva, in mucus, the gastric, intestinal or pancreatic juices, or in bile, in the digestive apparatus; or in the sweat, milk, tears, or lymph juices. Of these secretions, the milk, sweat, or saliva can serve to eliminate the poison entirely from the system ; the others can remove it only temporarily from the blood to place it under circumstances where it again meets conditions favorable for reabsorption into the blood by the vascular or lymphatic vessels. The laws which govern this intermediate circulation of poisons, as of the normal results of tissue change, are almost entirely unknown.

So far as the mere phenomena of diffusion are concerned, substances with low osmotic equivalents, such as soluble salts, will naturally tend to follow the path of the water in the organism, and hence first tend to appear in the fluid excretions, especially in the urine, where many of these bodies, such as iodide and ferrocyanide of potassium, are readily and rapidly detected after their entrance into the circulation. The rapidity of this excretion, as already suggested (page 65), can explain the non-appearance of symptoms of poisoning after the administration of the poison by the stomach, since the rapid removal by excretion of the substance from the blood prevents its accumulation in sufficient quantity to produce any evidence of its presence.

Section II.—Deposit of Drugs in the System through Assimilation.

The normal tissues are formed of solid and fluid elements, of which the former undergo a more gradual transformation than the latter. A foreign body, therefore, will have a tendency to remain longer in the system when it becomes a part of a formed tissue than when associated with one of the fluid elements; and if the part in which it is deposited is comparatively inactive in function (such as the bone), it may remain in the system indefinitely without producing any symptom as a result of its presence. It is only lately that examples of this form of assimilation of poisonous substances, which are always inorganic, have attracted any attention. In such instances it has been found that the inorganic principles in the animal economy may be replaced by other isomorphic elements. As examples of this may be mentioned the substitution of lead and baryta salts for the lime salts of the bones; the substitution for the phosphates of the isomorphic salts of arsenious acid; and probably also the deposits of copper in the various organs are governed by the same rules. As regards the ultimate fate of such foreign ingredients of the tissues it is probable that they are finally replaced by the normal constituents when the supply of the poison is interrupted.

Section III.—Chemical Alteration of Poisons in the Economy.

In the case of organic poisons, and probably also with the inorganic poisons in regard to substitutions in their acids and bases, changes in chemical constitution occur very frequently.

These changes may occur either at the point of absorption, especially when the poisons are administered through the digestive apparatus, only after entrance into

the blood, or finally in special organs; and it is conceivable that the same poison may undergo various
changes in different localities. These changes may convert a poisonous substance into one physiologically inert;
or, on the other hand, an active principle may be formed
from a substance in itself innoxious; or, again, the
modified form of a poison may produce symptoms differ-
ing from those due to its unaltered state.

Of the various possible modifications in constitution of
a poison, those attributable to changes at the point of
absorption are to be first examined, and, as already stated,
the digestive apparatus is most liable to produce such
changes as will convert a substance which is harmless
when administered subcutaneously into an active poison, or *vice versa*.[1] The character of many of these
changes may be indicated by their dependence on the
known characteristics of the locality of administration;
others can only be determined by the detection of their
products either in the excreta or in different tissues.

a. DISPLACEMENT OF ACIDS OR BASES IN SALTS.—
The double decomposition of two salts in mixed solutions,
by which precipitates are formed, or by which insoluble
bodies are thrown down while acids or bases are set free
in solution, may occur within the organism, governed by
the same laws as prevail elsewhere; and these changes by
rearrangement of elements may occur either at the point
of absorption (stomach or intestine), in the blood, the
tissue, or even in the excretory apparatus. Since soluble salts occur in all portions of the body, free acids in several localities, as in the gastric juice, and free alkalies in
nearly all the animal juices, the administration of any salt,
alkali, or acid offers an opportunity for the special changes
under consideration. And since usually all soluble salts

---

[1] In this connection it might be mentioned that poisonous substances may be produced by abnormal fermentations in the ordinary digestive processes, such as sulphuretted hydrogen, or ptomaines.

of a poisonous base or acid have the same action as the base or acid, such changes are consequently only of interest in this connection when some exception to this rule occurs, or when precipitates are produced in the rearrangement of molecules ; such changes may either lead to the elimination of the poisonous principle, may render it innoxious, or may mechanically produce functional disturbances by forming emboli.

b. FORMATION OF CHEMICAL COMPOUNDS WITH INGREDIENTS OF THE TISSUES AND EXCRETION UNDER MODIFIED FORMS.—The important discovery by Wöhler that benzoic acid, when introduced into the system, appears in the urine as hippuric acid, therefore united with glycin, was the first example of this kind. Other similar instances have only a physiological interest, but should serve, in the investigation of the ultimate fate of a poison, as a reminder of its possible destination, when no traces of the unaltered poison or of its oxidation products are to be met with in the excretions.

It is conceivable that to the formation of such compounds, either the poisonous action of the substance administered, or, possibly, its innoxiousness, may be due. No general rules for the determination of such conditions can be given.

c. DECOMPOSITION OF POISONS IN THE SYSTEM AND EXCRETION OF THE DECOMPOSITION PRODUCTS.—In general all organic, and certain inorganic, poisons are capable of decomposition, and the question arises whether the marked oxidizing property possessed by all the animal tissues can be concerned in the destruction of poisons. While formerly this was generally admitted without demonstration, latterly this possibility has been distrusted ; because, on the one hand, many poisons which were hitherto supposed to be so destroyed, can now readily be detected unaltered in the excretions, and, on the other, reliable physiological investigations have shown that apparently readily oxidizable substances, such as grape-sugar, are not thus consumed in the system.[1] Therefore, now, only when direct evidence of the products of such a decomposition is given, can this idea of the fate of a drug be entertained.

In this relation, however, the distinction must be drawn between complete and incomplete oxidation. Complete oxidation of organic substances is always accompanied by the formation of carbonic anhydride and water, and under special circumstances, by the liberation of nitrogen, and the formation of sulphuric and phosphoric acids ; the nitrogen appears never to be oxidized, or even to be completely isolated, but, on the contrary, remains associated with the other elements with which it was combined before the process of oxidation took place, and appears as ammonia or ammoniacal compounds in the excretions.

Of these different oxidation products, carbonic acid is the most readily detected. Carbonic acid formed by the oxidation of a poison can either be added to that already formed in expired air, thus increasing its proportion, or can be found as carbonates, especially alkaline carbonates, in the urine. Either of these possibilities can, however, only be established with any degree of reliability when extraordinarily large amounts of the poison are subjected to oxidation ; when therefore only small doses of the drug are dealt with, any attempt at such determination is useless. The method to be employed is similar to that employed in all estimations of the gases of respiration.[2] Large quantities of alkaline carbonates are indicated in the urine by an abnormal alkaline reaction : for their careful qualitative or quantitative analysis see Hoppe-Seyler, Analyse, p. 256.

The investigation as to the incomplete oxidation of poisons will depend upon the detection of special oxida-

[1] Ludwig and Scheremetjewski. Sächs. Acad. Sitzgsber, 1869. 154.

[2] For simple methods see Kowalewski and Sanders-Ezn. (Sächs. Acad. Sitzgsbr. 1866, iii. ; 1867, 68 ; 1869, 154) and Röhry and Zuntz. (Arch. f. d. Ges. Physiol. iv. 57).

tion products, such as of aldehyde for alcohol, benzol or
phenol for acetic acid, etc.   Among the normal products
of incomplete oxidation, whose increase is to be looked
for, oxalic acid deserves notice.   As already stated, these
processes of oxidation may lead to the removal of the
poisonous principle from the blood, or when the decomposition products are themselves poisons, to new symptoms of poisoning.   In the latter case the symptoms can
only be attributed to such products after the latter have
been proved to be present.

    d. DECOMPOSITION OF POISONOUS SUBSTANCES AND
EXCRETION OF THEIR PRODUCTS.—The animal organism
is rich in ferments which are capable of causing hydro-
lytic decomposition of various substances (*i. e.*, their
division with the addition of water); and there is no
doubt that poisons, like many of the normal constituents
of the body, may be subjected to such changes, and the
destruction of the poisonous properties of the drug, or
the creation of new poisonous principles, be thereby produced.   Starting-points for such studies can, however, be
obtained only from the knowledge of the chemical properties of the drug in question.

    With the outlines given above all the possible changes
of the poison in the economy are not exhausted.   And
it cannot even be said that a complete knowledge of the
fate of a poison is possessed when its ultimate products
and the paths by which they leave the system are known.
In the first place all intermediate changes should be
studied, and in the second the question will arise as to
the location in which these phases are met with, and
especially the way in which the particles of the poison
produce their characteristic action.
    But since we possess but little knowledge as to the
changes in the normal constituents of the body, we can
naturally not expect to obtain these data for poisons, even
though the latter can the more readily of the two be followed through the system; nevertheless careful phar-

macological studies of these points will undoubtedly ultimately be of the greatest assistance to physiology. But as yet, even in the case of substances most readily detected, such as the poisonous metals, we know little more than that they are excreted through the kidneys in some unknown form.

IV.

EXPLANATION OF THE SYMPTOMS PRODUCED BY POISONS.

THE manifestations of the action of a poison may belong to one of two classes : *First.* Those which can be attributed to local action on the absorbing surfaces, and which are either confined to the absorbing surface, or which may be capable of producing secondary general disturbances of function from sympathy. *Second.* General manifestations in the strictest sense, and which are dependent upon the presence of the poison in the blood.

No fundamental points of contrast can be made between the local and general action, since, as already indicated, the general is but the sum total of all the local effects of the poison produced by its entrance into the blood and its simultaneous action on all the organs.[1] When it is remembered, however, that in the local application the poison comes in contact with the tissues in a very different manner from when introduced into the blood, a partial separation of the two groups of phenomena is rendered possible. In local applications, the

----

[1] For a long time it was believed that poisons, especially the narcotics, were carried throughout the system by the agency of the nervous system. This supposition can, however, be disproved by the following experiment: The hind leg of a frog is partially separated from the body, leaving only the sciatic nerve intact, and immersed in a solution of strychnine, without any symptoms being produced ; if, however, instead of the nerve, the bloodvessels are left intact, symptoms of poisoning rapidly ensue.

8

poison acts exclusively, or at least in the first place, on the superficial tissues of the organ, the degree of action depending largely on the concentration of the solution; while after introduction into the blood it acts gradually on all portions of the organ, perhaps weakened by dilution by the blood, or by the chemical alterations to which it has been subjected. Thus it may happen, that a poison will act in various ways on a single organ, according as it is directly applied, or carried to it by the circulation; in one case, possibly, being extremely energetic, in the other, completely inactive. The only method of determining to which of these modes of action the symptoms are due, if not self-evident, lies in change of the mode of administration.

In all cases, the symptoms of poisoning, especially in the warm-blooded animals, furnish a complex picture which can only be explained by a systematic series of experiments; it is the object of the following pages to give some general rules whereby such experiments can be conducted.

After an answer to these questions has been found, and the organ or organs determined on which the poison, after introduction into the circulation, principally acts, and the character of the functional disturbance established, it then remains to discover to what physical or chemical peculiarity of the poison its action is due. The reply to this question, which has as yet been reached in but few cases, will at the same time explain the reason of the localization of the poison in any special organ.

A third important point of inquiry arising in the careful study of the action of a drug concerns the mode of recovery after poisoning; that is, the manner in which the system, and especially the affected organs, free themselves from the poison and regain their normal condition.

[After the administration of the poison the attempt should first be made to obtain an idea of the general action of the poison, so as to facilitate the localization and analysis of the separate symptoms. The animals, therefore, after receiving a dose of the poison, are allowed to move about (frogs may be placed under a glass bell-jar), and notes made on the different symptoms as they appear, with reference to the time elapsing between the administration of the poison and the advent of the various symptoms. It should be noted whether the animal is restless and inclined to move about, and whether his movements appear normal, or unsteady, weak, spasmodic, etc ; or whether he seems disinclined to move, and if so, if the disinclination depends upon loss of power, pain, or somnolence. Notes should also be made as to occurrence of vomiting, purging, or micturition, salivation, lachrymation, state of the pupil, etc., as perhaps furnishing guides to the detection of the specific action of the drug. The heart beats and respiration should be counted, and any change in character noted, and if the drug prove poisonous, and the dose fatal, it should be observed whether respiration ceases while the heart still continues to beat, or *vice versa*. As soon as possible after death, the animal should be opened, and the condition of the heart, whether pulsating or not, or whether stopped in systole or diastole, noticed ; it should also be determined whether the cardiac and voluntary muscles and motor nerves preserve their irritability to various stimuli, whether the intestines are in active peristalsis or not, and the amount and color of the blood in the lungs and right side of the heart.

If after waiting some time no decisive results of the poison appear, the dose should be repeated at intervals until the drug is proved to be innocuous, or until some effects have been produced. The mode of administration also should be changed in either case for the reasons already given.

If the drug be poisonous, the next question is to determine the smallest dose capable of causing death. To accomplish this an animal is weighed, a small dose injected, and after waiting a short time, the dose repeated until death is produced: it is then calculated how much of the poison per gramme weight of the animal was necessary to prove fatal.   Another animal is then weighed, and a single dose of the poison, which should be a little less per gramme weight than the total amount obtained in the previous experiment, injected at once, a smaller amount being selected so as to allow for possible excretion during the intervals of the first experiment.   If this amount prove fatal, a still smaller proportionate dose is given to another animal, until the smallest possible dose capable of causing death has been determined.]

The study of the symptoms of poisoning will naturally start with the one which is most marked.   In many cases therefore a departure from the order of investigation here followed will be advisable.

Section 1.—Action on the Circulatory Apparatus.

Direct examination of the heart will furnish a means of determining the frequency, strength, and rhythm of its pulsations.   The first of these points, and to a certain degree both the others, can be ascertained in the larger animals and in man, without any operation, through palpation of the cardiac impulse or pulse, or by auscultation of the heart sounds.   In animals the frequency of pulsation may be rendered evident by inserting a long needle through the chest wall into the heart, when its oscillations, coinciding with the pulse, may either be counted, allowed to register on a revolving drum, or to strike a bell, thus serving for demonstration to large audiences.   In frogs it is necessary to expose the heart, and this is readily accomplished by removing with scissors a triangular piece of the skin and bony wall of the thorax over the heart so as to expose it completely ; it is better to leave the pericardium uninjured.   In warm-blooded animals it is very difficult to expose the heart without injuring the pleura, and even if the operation should succeed it will not freely expose the heart ; when, therefore, a careful inspection of the heart is necessary in warm-blooded animals it is better to freely open the thorax and maintain artificial respiration. The heart is then readily inspected during the pauses between inspiration when the contracting lungs uncover it.

More satisfactory results as to the rapidity of the heart's pulsations and its relation to various conditions of the vascular system are obtained by the graphic method. This is accomplished either by the Marey or Pond sphygmograph applied over an artery, the cardiograph over the cardiac impulse, or the Ludwig or Fick manometer connected with an artery and serving to register the oscillations of blood-pressure on the kymographion.

[The best form of cardiograph is that of Marey and Chauveau as modified by Sanderson (Fig. 16). It

Fig. 16.

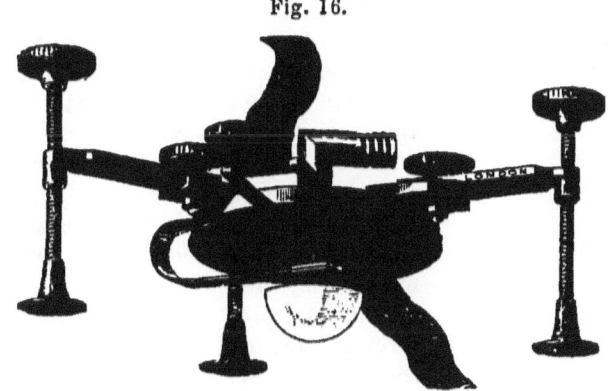

Sanderson's cardiograph.

consists of a shallow disk-shaped box of metal, the top of which is closed by a rubber membrane, while its interior communicates with a small rectangular tube inserted in the bottom. Fastened to the bottom of the box is a flat steel spring, bent on itself twice at right angles so that its free extremity, which is perforated by a steel screw carrying a large button is opposite the centre of the membrane, where the latter is protected by a thin disk of metal.

The box or tympanum is also provided with three

8*

Fig. 17.

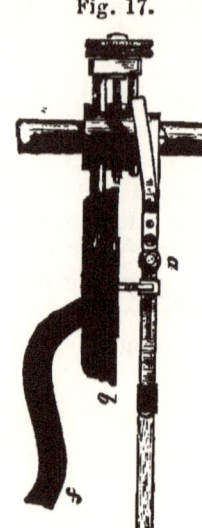

levelling screws and tapes to fasten it to the body. When an observation is to be made, the instrument is bound to the chest so that the knob on the head of the centre screw lies exactly over the point of cardiac impulse. The rectangular tube is then connected by means of rubber tubing with Marey's tympanum and lever (Fig. 17), an instrument constructed on the same principle as the cardiograph, with the exception that the movements of its rubber membrane are communicated to a lever by which they can be recorded on a revolving drum covered with smoked paper. When now the levelling screws are so adjusted that the head of the screw passing through the spring presses on the chest wall while its point is in contact with the rubber membrane, every elevation in the chest wall will produce a corresponding depression in the rubber membrane, thus diminishing the size of the cavity of the tympanum and forcing some of its contained air into the second tympanum whose membrane will be proportionately expanded, its motion being magnified by the lever resting on it.

When, therefore, a tracing is taken of the cardiac impulse, it will be found that the systole of the heart is always indicated by a sudden ascent of the lever, and the diastole by a gradual fall. With the apparatus so arranged studies may be made on

Fig. 17.—Marey's tympanum and lever.

the effect of drugs not only on the absolute rate of pulsation of the heart, but also on the relative duration of the ventricular systole and diastole. Care must be taken, however, not to attribute all changes from this normal simple curve to toxic action; for if the cardiograph becomes at all shifted, so that the button lies to one side of the impulse, the character of the tracing will be entirely changed. For here, while we still obtain a sharp ascent of the lever corresponding to the ventricular systole, the rise is immediately followed by a sudden fall, due to the recession of the chest-wall in the production of the so-called negative impulse.

The purpose of the sphygmograph is to represent graphically the succession of expansions and contractions which occur in the artery consequent to each cardiac systole; its use and construction are so well known as to require no description here.]

The sphygmograph has the advantage that it requires no operation for its use, and can therefore be employed in experiments made on man; and both the sphygmograph and cardiograph, on careful analysis of their curves, give information not only as to the frequency of the heart's pulsations but also as to the duration and characteristics of the single periods of each cardiac contraction.

Although these instruments are valuable as furnishing a means of studying the different times of the heart's beat, yet the force of the beats can only be measured with manometers, either the simple hæmodynamometer, or recording manometer (kymographion).

The kymographion serves to measure the blood-pressure in an artery, besides enabling the frequency of pulsation to be counted. The blood-pressure is dependent not only on the strength and frequency of the heart's action, but also on the condition of the general vascular system; every contraction in the arterial system, whether general or confined to a single extended portion of the body, produces an increase in arterial blood tension. So the results obtained by kymographic measurements

cannot be attributed directly to the heart, but will depend
upon the condition of the arterial system at large.

MODE OF CONDUCTING AN EXPERIMENT ON BLOOD-
PRESSURE.—[The tension of blood within the arterial
system may be measured in various ways; either by the
primitive method of Hale, who simply connected a long
vertical glass tube with an artery and noted the height
to which the column of blood rose, or by the cardiometer
of Poiseuille, or the hæmodynamometer of Bernard; the
improved manometers of Ludwig and Fick marked such
an advance in this line of investigation that one or the
other of their instruments is now invariably used.

Ludwig's manometer (Fig. 18) consists of two communicating glass tubes m and m', partially filled with
distilled mercury, inserted into a block of steel in which
a cavity is hollowed out which communicates with the
interior of the two glass tubes; below is an opening
closed by a steel screw which can be removed when the
instrument requires cleaning. To the left the glass tube
communicates with a flexible leaden tube t, a stop-cock c
intervening, which is connected with the artery in which
it is desired to measure the blood-pressure. The mercury
in the arm m' of the manometer bears on its surface an
ivory float connected with a long slender steel rod carrying the pen p, for recording the movements of the column
of mercury on the revolving surface r. At pb is a box
containing a pressure-bottle, filled with a saturated solution of sodium carbonate, which can be elevated or depressed at pleasure.

The entire system of tubes between the surface of the
mercury in the tube m of the manometer and the artery
and pressure-bottle being filled with sodium carbonate,
by which the blood is prevented from clotting, the clamp
c'' being closed and the stop-cock c and the clamp on the
artery opened, the column of mercury in the distal arm of
the manometer rises to a point which indicates the blood-
pressure, while its oscillations indicate the rapidity of
the heart's pulsations. As, however, these variations in

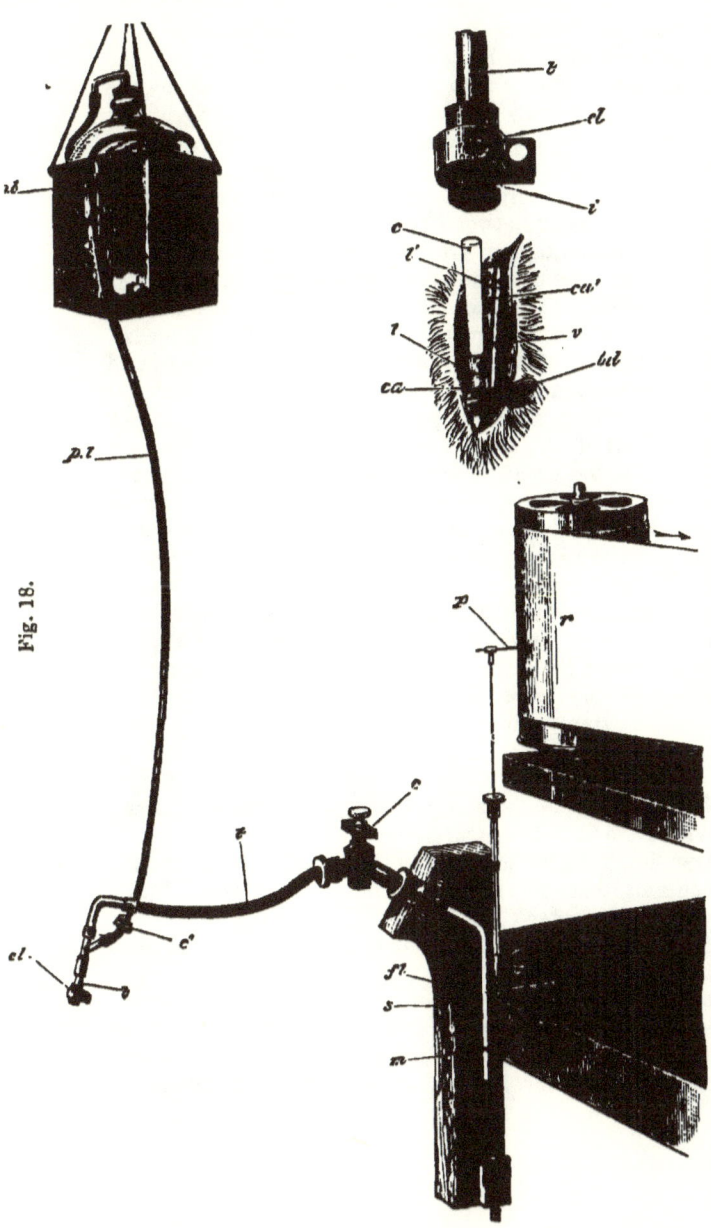

Fig. 18.

Ludwig's manometer arranged for investigating blood-pressure. (From Foster's Physiology.) At the upper right hand corner is seen, on an enlarged scale, the carotid artery, clamped by the forceps *bd*, with the vagus nerve *v* lying by its side. The artery has been ligated at *l'*, and the glass canula *ca* introduced into the artery between the ligature and the forceps; the shrunken artery is seen on the distal side of the canula at *ca'*. Above is seen the clamp *cl* of the arterial tube; *f* is a piece of heavy rubber tube, surrounded by the clamp *cl*, into which the arterial canula and manometer tube are inserted and made fast by tightening the screw.

pressure produced by the heart's pulsations are in most animals too rapid to be counted by the unaided eye, the movements of the column of mercury are recorded by the pen on the moving surface, the vertical height of the tracing thus produced indicating the blood-pressure while the distance between the breaks in this line, the rate of motion of the recording surface being known, will give the time elapsing between each cardiac revolution.

The disadvantage of all forms of mercurial manometer is that the inertia of the mercury is so great that it does not give the true form or extent of the variations in blood pressure which accompany each beat of the heart. Hence the use of the mercurial manometer is restricted to the investigation of those changes in mean arterial pressure which are not interfered with by the proper oscillations of the instrument, that is, when they do not recur with too great rapidity; while it must be remembered that the curve is not that produced by the actual movement of the arterial column, but of the mercury. For while the artery expands suddenly the mercury rises slowly, and before the latter has attained its highest point the artery will have collapsed and the falling column of mercury will meet an expanding force in the artery. Hence, when the heart's beats are very rapid, the oscillations of the mercurial column will be relatively too small in extent, while when the heart beats slowly, they will be exaggerated.

While, therefore, the mercurial manometer answers perfectly well for the study of the mean arterial pressure and change in the number of cardiac pulsations, exact information as to the changes in the arterial system during each pulsation can only be obtained by the use of Fick's spring manometer (Fig. 19). This consists essentially of a C-shaped hollow spring of thin metal a, filled with alcohol, and connected, by a leaden tube attached to c and filled with sodium carbonate solution, with the artery.

Each variation in the arterial pressure will cause an expansion or contraction of the spring, whose movements, magnified by a lever l, may be recorded on the revolving drum; before use, this manometer must be graduated by comparison with a mercurial manometer. In ordinary pharmacological work the mercurial manometer is sufficiently accurate.

Fig. 19.

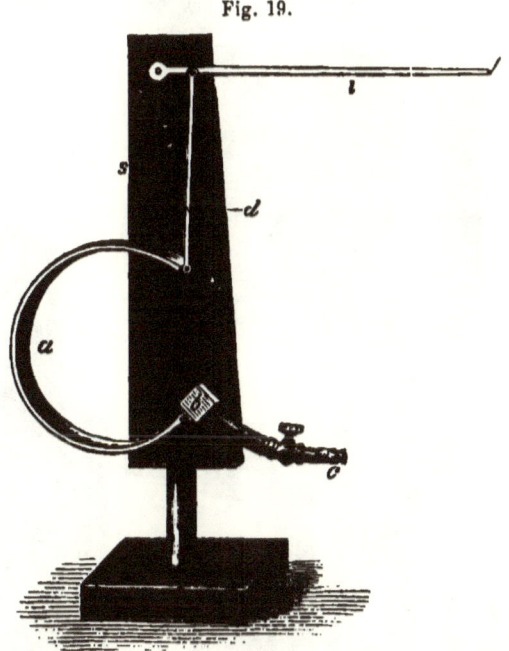

Fick's spring manometer. (From Foster's Physiology.)

The present important system of recording known as the "graphic method" started with the simple revolving drum of Ludwig, and while to him belongs the credit of having first introduced this method into physiological experimentation, to him also do we owe the majority of the modern improvements in recording apparatus as now used.

The kymographions in general use in the physiological laboratory for studies of blood-pressure are of two kinds:

Fig. 20.

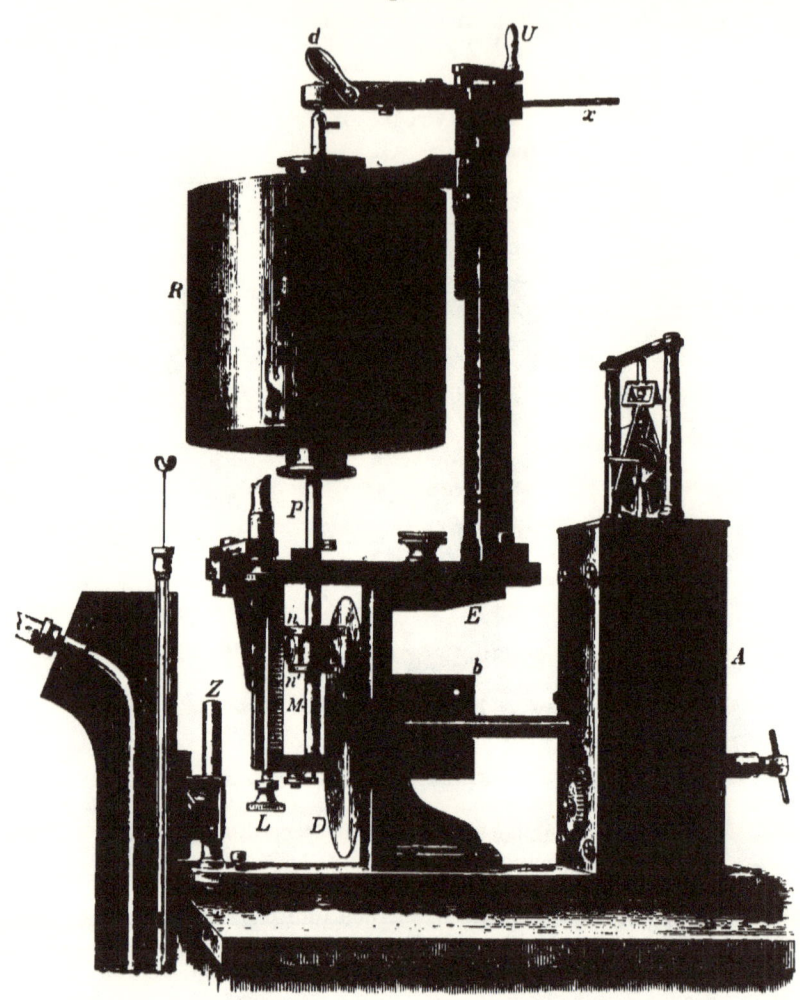

Ludwig's kymographion. (From Cyon's Methodik.)

in one the manometer pen traces its movements on a revolving drum covered with smoked paper ; in the other,
the pen contains ink and describes its curves on a uni-
formly moving surface of unglazed paper.

The first of these instruments is represented in Fig. 20.
The clock-work contained in the case *A* sets the revolving
disk *D* in motion and transfers its movements by means
of the friction rollers *n* and *n'* to the metal drum *R*. The
position of this drum can be changed at will by means of
the screw *U*, while its rate of revolution may be graduated by the screw *L* (it of course moving more rapidly
when the friction rollers are at the circumference of the
disk than when at its centre), and by adjusting the clock-
work. Two drums are usually supplied with each instrument, so that when one is filled it can be removed by
raising the clamp *d* and the other substituted. At *Z* is
seen the mercurial manometer.

To obtain information as to the mean arterial pressure
all that is necessary is to allow the manometer to write
on the revolving surface, when the height of the curve
above the abscissa will give the desired information ; but
studies of changes in the pulse as well as of the times
of blood-pressure variations, require that the rate of
motion should also be recorded on the drum. Nearly all
forms of such time-recorders, when employed for this purpose, are modifications of the simple electro-magnet, first
used by Dr. Locke, of Cincinnati ; nearly any form of
electro-magnet, such as a telegraph sounder, can be made
to answer this purpose by attaching a light lever to the
keeper, and allowing it to record its motions on the drum ;
for use as a time-marker, the current through the electro-
magnet must of course be interrupted at regular, known
intervals by an electric metronome, or by any of the
various automatic current-interrupters. For description
of such instruments, reference must be made to some of
the various hand-books for the physiological laboratory.[1]

---

[1] A simple form of automatic current interrupter, employed by the
Translator, is described in the Medical News, Sept. 30, 1882, p. 370.

9

The large kymographion with continuous roll of paper is represented in Fig. 21.

Fig. 21.

Large kymographion, for continuous tracings. (From Foster's Physiology.) The clock-work machinery, some of the details of which are seen, unrolls the paper from the roll C, carries it smoothly over the cylinder B, and then winds it up into the roll A. Two electro-magnetic markers are seen in the position in which they record their movements on the paper as it travels over B. The manometer, or any other recording instrument used, can be fixed either in the notch in front of B or in any other position that may be desired.

DETAILS OF BLOOD-PRESSURE EXPERIMENTS.—Blood-pressure experiments may be made on dogs, cats, or rabbits, the artery selected being either the carotid or femoral.

When the small kymographion is used, both drums should be smoothly covered, before commencing the operation, with strips of glazed paper of the width of the drum and a little longer than the drum's circumference, and the overlapping ends fastened with mucilage, care being taken that the paper is not pasted to the drum, or it will be impossible to remove the paper after the experiment without damaging the tracings. The drums are then smoked by revolving them rapidly in the flame of a coal-oil lamp, or in that of burning camphor, until they are covered with a uniform coating of soot; it is better not to coat the drums too heavily, or the friction will be so great that a clear tracing cannot be obtained. One drum is then placed in position and the clock-work wound up. When the large kymographion is used, the paper is of course not smoked, the tracings being made with glass pens filled with a one per cent. aniline solution.

The animal is then fastened on its back and a canula inserted in the usual manner into the carotid or femoral artery, care being taken in the former case not to injure the important cardiac nerves lying in the carotid sheath ; a second canula, fitting the nozzle of the injection syringe, is then inserted into the jugular vein.

The test should then be made, by closing the end of the tube t with the finger and rapidly compressing it at another point with the fingers of the other hand, as to whether the manometer pen rides freely on the surface of the mercury ; the observance of this precaution will save a great deal of annoyance in the course of the experiment from clogging of the mercury around the float ; no experiment should ever be commenced until all the apparatus has been tested and found in perfect working order.

The manometer pen is held in contact with the revolving surface by means of a little lead plummet attached by a thread to the brass rod seen at x on the top of the kymographion ; this rod must of course be moved until its extremity is vertically above the manometer pen. After having seen that the current-interrupter and electro-magnetic time-marker are in good working order, the latter is so adjusted as to write on the drum vertically under the manometer pen, and at a height corresponding with the level of the latter when both columns of mercury are of the same height ; the line then described by the time-marker will serve as an abscissa, or line from which the blood-pressure is measured.

The pens should barely touch the smoked surface so as to interfere as little as possible by friction with the motion of the drum.

At the outset of the experiment the drum should be screwed up until it occupies its highest possible position.

When, as in the diagram, the tube which contains the soda solution ends in a simple clamp, the arterial canula must be first completely filled with soda solution by means of a pipette, and then the clamp c'' opened until all the air is expelled from the tubes, and while the soda solution is still running, taking care that none flows into the wound, the end of the tube t is to be slipped over the arterial canula and the clamp cl fastened : by this means the entire system of tubes from the artery to the surface of the mercury is completely filled with soda solution, it being very important that no air bubbles are present.   The stop c is now opened and the pressure bottle raised until the height of the distal column of mer-cury indicates a pressure a *little less* than the expected blood pressure : it is important that the tubes should be filled with fluid under pressure, otherwise the blood, when the arterial canula is opened, would flow into the tubes and so tend to clot, while it is also important that this pressure should not be greater than that to which the blood is subjected in the vessels, or the soda solution would flow into the arteries.

After this point has been attended to, the clamp c'' is closed and *then* the clip or slip-noose on the artery loosened : the column of mercury will now rapidly rise a short distance and be the seat of rapid oscillations depend-ing on the heart's beats, while larger curves of movement will be superimposed upon these by the respiratory movements.

The canula inserted into the vein is then filled by means of a pipette with the solution to be injected so as to exclude air, and the nozzle of the filled injecting syringe inserted and bound fast.

If, during this procedure, it is found that the arterial pressure has remained constant, the drum may be started, the time of starting being written on the drum ; after fifteen or twenty seconds have passed the clip or noose can be removed from the jugular vein and the dose of the poison injected, the beginning and end of the injection and the amount injected being marked on the drum. After the drum has made a complete revolution, it may be depressed by means of the screw u, so as to furnish a fresh surface for the tracing. If no effect followed the first dose a second may now be given, and the doses may be repeated at the desired intervals. When one drum is filled with tracings, the clock-work is stopped, the time being written on the drum, the drum removed and a fresh one substituted; in this way tracings may be taken for several hours in succession.

The paper is removed from the drum by cutting with a sharp knife and the tracing fixed by passing the paper through a bath of shellac dissolved in alcohol, and then suspending the papers until dry.

Occasionally, during the course of an experiment, a clot will form in the canula and prevent the registration of the pulse and pressure curves. When this occurs, if the above described forms of canulæ are used, a clip must be placed on the artery and the canula removed and washed out by allowing the soda solution to flow through it, though sometimes the clot may be displaced by undoing the clamp cl, and picking out the clot with a bristle or straw; before recommencing the experiment, the apparatus must again be arranged as before.

To prevent this inconvenience and loss of time in removing canulæ to displace blood-clots, it is a great advantage to employ between the arterial canula and the clamp on the tube t, a canula similar to the one represented in Fig. 22. This consists of a T-tube of silver, of which the shorter arm is bound into the tube t while the end a is inserted into the arterial canula; the end b is closed by a plug; and the line shown in section represents a partition. When now a clot forms in the arterial canula or in the tubes, a clip is placed on the artery, and the plug b withdrawn; the soda solution

9*

then running in the direction indicated by the arrows usually serves to wash out any clots.

Fig. 22.

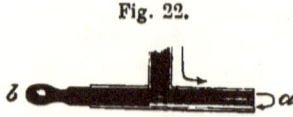

In order to facilitate the deduction and comparison of results obtained in the above method, it is necessary to tabulate the curves. Several plans may be pursued; the simplest, and one sufficiently accurate for all general purposes, is as follows: The line described by the second-marker is divided off into units of fifteen seconds, and perpendiculars, intersecting the pulse-curve, erected at the commencement of each of these spaces; each break in the manometer tracing represents a single contraction of the heart, and the number of these breaks between two perpendiculars will give the number of heart-beats in fifteen seconds. The mean blood-pressure in any fifteen seconds may be determined by measuring the distance between the time-line, which serves as the abscissa, and the highest and lowest points on the manometer tracing between two perpendiculars and adding them together: the two measurements are made because of course the tracing simply gives the height of ascent in one arm of the manometer, while since the mercury falls to a corresponding extent in the opposite arm, the blood in the vessels is always sustaining a pressure of a column of mercury of double the height of the tracing above the abscissa. The results thus obtained are then tabulated as follows: it being noticed that in this instance the time is divided into periods of ten seconds, while the lower line represents the abscissa. The tracing is to be counted from right to left. (Fig. 23.)

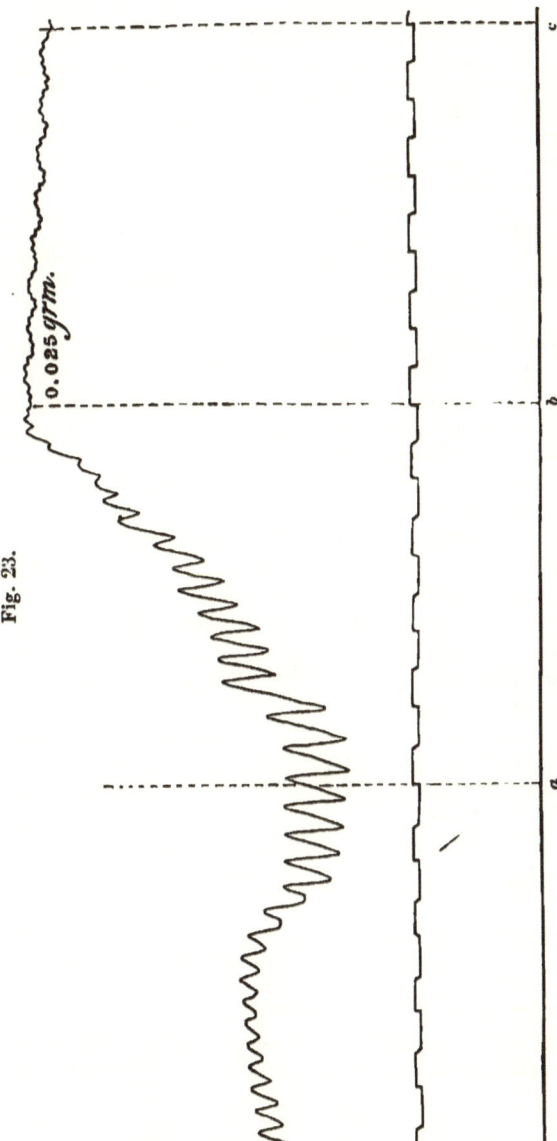

Fig. 23.

Blood-pressure tracing from an experiment on a rabbit with the alkaloid of sanguinaria Canadensis; slightly reduced. The lower line is the abscissa; the breaks in the second line represent seconds; the upper line is described by the manometer pen. The ordinates a, b, c, are erected to cut the pulse line at intervals of 10 seconds; the respiratory curves may be seen at the commencement of the tracing.

Experiment.   No.   Date.

Large male rabbit; weight   grms.   Canula in the left carotid; canula in the jugular vein.   Animal otherwise uninjured.

Time.	Time after injection.	Blood pressure in mm. of mercury.	Pulse in 15''.	Remarks.
12 : 0 : 15	...	135	42	0.025 grm. Sanguinarina
12 : 0 : 25	10	95	18	sulph. injected into jugu-
12 : 0 : 35	20 sec.	65	16	lar towards the heart.

Or the results may be plotted to scale on profile millimeter paper.

When it is found that both pulse and pressure are remaining constant, it is of course not necessary to tabulate long rows of figures which show no change; account should, however, always be kept of the time elapsing between different tracings.

The respirations may also often be counted in these experiments, as seen in the respiratory waves in the manometer tracing.

It is not claimed that the method above given, though sufficiently correct for general purposes, furnishes a means of accurately measuring the blood-pressure.   The mean pressure is most accurately obtained by means of a planimeter.   Another method is to determine the square superficies of the irregular figure bounded by the abscissa, the curve and the two perpendiculars, and then divide it by the length of the abscissa.   The size of the figure may be determined by placing over it a piece of tracing paper or glass ruled in square millimeters, and counting the number of squares contained in it.   In very accurate experiments, the weight of the column of sodium carbonate in the manometer should also be deducted; when a solution of sp. gr. 1015 is used, this fraction amounts to about $\frac{1}{27}$ of the whole.]

The action of a poison on the heart may be manifested either in alterations in its frequency of pulsation (acceleration, retardation, or arrest in systole or diastole), in alterations in the strength of contraction, in disturbances of the normal cardiac rhythm (for instance, each ventricular systole may be preceded by two auricular contractions, etc.), or in various changes from the normal blood-pressure. The majority of heart poisons produce first acceleration, then retardation, and finally irregularity, loss of power and arrest of the cardiac pulsations.

CAUSES OF THE CHANGES IN THE CIRCULATORY MECHANISM.[1]

[Before we attempt to discover the organ or organs to whose functional disturbance the results produced by the drug on the circulatory mechanism are due, it may be well to first give a short outline of the possible causes of variation in blood-pressure and pulse-rate.

In the first place, it must be remembered that the blood-pressure depends not only on the amount of blood pumped into the arteries, but also on the amount of blood which flows in the same time into the veins: Consequently the blood-pressure may be raised, 1, by the heart beating more quickly; 2, by a larger amount of blood being pumped into the aorta by each beat; 3, by preventing the escape of arterial blood into the veins from contraction of the small arteries. And blood-pressure may be lowered by, 1, a slow rate of cardiac pulsation; 2, by imperfect ventricular contraction, whereby only a small amount of blood is pumped into the aorta at each pulsation; 3, by relaxation of the small arteries, whereby the escape of blood into the veins is facilitated; 4, by obstructed pulmonary circulation.

One or the other, or may be several, of these physical causes will always lie at the bottom of all variations in blood-pressure. But as we know that both the rate of

---

[1] In the preparation of this section, the Translator has made free use of Dr. Lauder-Brunton's lectures on "The Action of Drugs on the Circulation."

the heart's contraction and the condition of the small
arteries depend upon impulses coming from the nervous
system, we will find that circulatory disturbances produced by drugs are usually due to some interference with
the regulating nervous mechanism, though there exists
the possibility of direct action on the muscular tissue of
the heart or of the arteries.

The nervous regulating mechanisms of the heart are
found in the cardiac ganglia, the inhibitory nerves, the
accelerating nerves, and the vaso-motor system.

1. THE CARDIAC GANGLIA.—That the heart contains
within itself the conditions necessary for its rhythmical
movement is a fact whose knowledge dates from the time
of Galen, but that the explanation of this phenomenon lies
in the function of automatic nervous centres situated in
the walls of the heart, was first pointed out by Remak.
These cardiac ganglia are three in number, and are of
different functions ; two are motor ganglia, one an inhibitory ganglion.   The motor ganglia are the ganglion
of Remak, situated at the opening of the inferior vena
cava, and the ganglion of Bidder situated in the left
auriculo-ventricular septum; the inhibitory ganglion of
the heart, that of Ludwig, is situated in the inter-auricular
septum.   These ganglia are not only automatic in function, but are also under the control of, or capable of being
modified by impressions coming
from the central nervous system,
and by varying conditions in the
temperature and chemical composition of the blood.

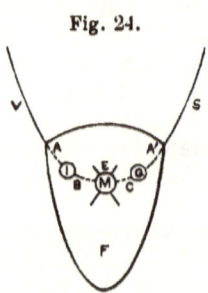

Fig. 24.

Diagram of the hypotheti-
cal nervous apparatus of the
heart.

The elaborate studies of Schmiedeberg on heart poisons have rendered necessary the assumption of
a still more complicated system of
intrinsic cardiac nervous system;
this hypothetical nervous apparatus is represented in the diagram,
Fig. 24.   (Brunton.)

The motor ganglion M maintains

the rhythmical contraction of the muscular fibres of the heart with which it is in connection through the fibres E. This motor ganglion is connected by an intermediate apparatus with the inhibitory ganglion I, and the latter by the fibres A with the centrifugal inhibitory influence passing through the pneumogastric nerves ; by means of this mechanism the motor impulses generated by the ganglion M may be arrested or retarded. The ganglion M is further connected by an analogous apparatus with the accelerator ganglion Q, and the latter by the fibres A' with the accelerator nerves coming from the medulla and sympathetic nervous system.

It is possible for poisons to produce cardiac disturbance by interference with the functions of any one part of this apparatus ; the methods for determining what structures are affected will be subsequently given.

2. THE CARDIAC INHIBITORY NERVES.—The inhibitory nerves of the heart arise in the cardio-inhibitory centre in the floor of the fourth ventricle and reach the heart through the pneumogastric nerves ; when they are irritated the pulse is slowed, or the heart may be arrested in diastole. In man and the dog, cat and rabbit, they are in constant action, and when divided, the heart beats more rapidly; the increase in the pulse after section or paralysis of the vagi is more marked in the dog than in the cat or rabbit.

Drugs may render the heart's pulsation slow by, 1, *direct irritation* of these inhibitory fibres either, *a*, at their origin in the medulla, *b*, in their path through the vagi, or, *c*, in their terminal fibres in the heart.

Or 2, the pneumogastrics may be *indirectly* irritated through the action of the drug on other parts, producing, *a*, increased blood-pressure, or, *b*, accumulation of carbonic acid in the blood.

3. The inhibitory nerves may be *reflexly* irritated through stimulation of sensory nerves, irritation of the intestines, of the sympathetic, of the depressor nerve, or of the vagus of the opposite side.

On the other hand, drugs may paralyze any point in

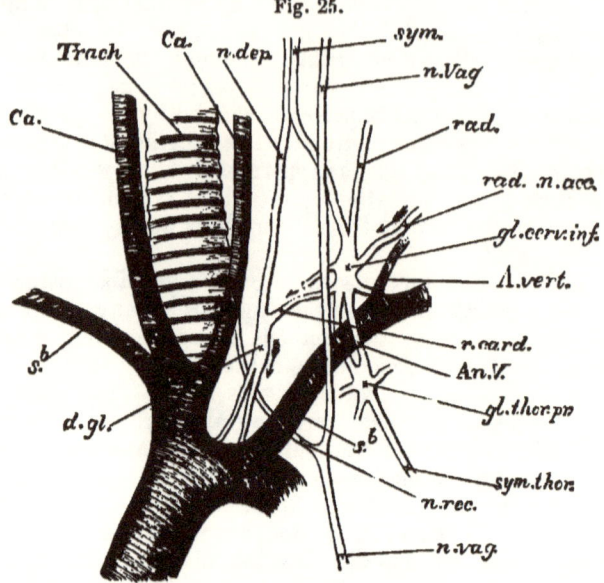

the course of the inhibitory fibres, and thus quicken the heart.

Fig. 25.

Diagram of the last cervical and first thoracic ganglia in the rabbit. (From Foster's Physiology.) *Trach.* Trachea. *Ca.* Carotid artery. *Sb.* Subclavian artery. *n. vag.* Vagus trunk. *n. rec.* Recurrent laryngeal. *sym.* Cervical Sympathetic ending in inferior cervical ganglia, *gl. cerv inf.* Two roots of the ganglion are shown, *rad.*, the lower of the two accompanying the vertebral artery, *A. vert,* being the one generally possessing accelerator properties. *gl. thor. pr.*, the first thoracic ganglion. Its two branches communicating with the cervical ganglion surround the subclavian artery forming the annulus of Vieussens. *Sym. thor.* The thoracic sympathetic chain. *n. dep.* Depressor nerve. This is joined in its course by a branch from the lower cervical ganglion, there being a small ganglion at their junction, from which proceed nerves to form a plexus over the arch of the aorta. It is this branch from the lower cervical ganglion which possesses accelerator properties—hence the course of the accelerator fibres is indicated in the figure by the arrows.

3. ACCELERATOR NERVES.—The accelerator nerves arise in the medulla oblongata, pass down the cervical portion of the spinal cord and join the last cervical and

first dorsal ganglia, and thence to the accelerator ganglion of the heart; their distribution is somewhat different in the dog and the rabbit. (See Figs. 25 and 26.) Unlike the inhibitory nerves of the heart, they are not in con-

Fig. 26.

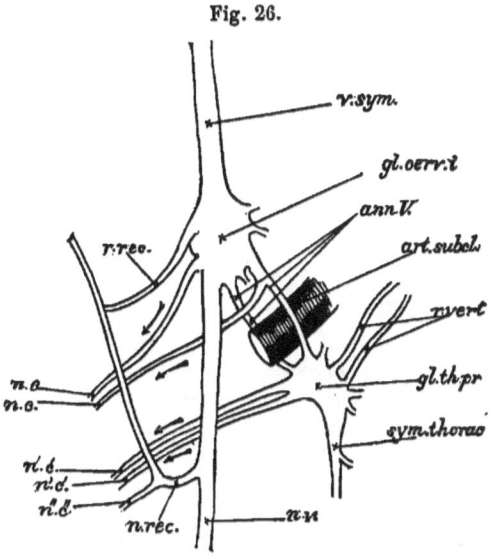

Diagram of the last cervical and first thoracic ganglia in the dog. (From Foster's Physiology.) *v. sym.* The united vagus and cervical sympathetic nerves. *gl. cerv. i.* The inferior cervical ganglion. *n. v.* Continuation of trunk of vagus. *ann V.* The two branches forming the annulus of Vieussens around the subclavian artery, *art. subcl.*, and joining *gl. th. pr.*, the first thoracic or stellate ganglion (the branch running in front of the artery is considered by Schmiedeberg to be an especial channel of accelerator fibres). *Sym. thorac.* The sympathetic trunk in the thorax. *r. vert.* Communicating branches from the cervical nerves running alongside the vertebral artery, the rami vertebrales. *n. rec.* The recurrent laryngeal. *n. c.* Cardiac branches from the lower cervical ganglion, accelerator nerves of Schmiedeberg. *n'. c'.* Cardiac branch from first thoracic ganglion, accelerator nerves of Cyon. *n''. c''.* Cardiac branch from recurrent nerve. *n. rec.* Branch from lower cervical ganglion to the recurrent nerve, often containing accelerator fibres.

stant action; hence their section or paralysis is not followed by slowing of the heart. Drugs may produce
10

an increased rate of pulsation by direct stimulation of these nerves either at their origin, in their course or at their termination in the heart, or indirectly by producing a diminished blood-pressure. The influence of drugs on the accelerator apparatus of the heart has not been as thoroughly worked out as in the case of the inhibitory mechanisms.

4. VASO-MOTOR SYSTEM.—The normal tonicity of the bloodvessels is maintained by the vaso-motor centre located in the floor of the fourth ventricle of the brain.

Blood-pressure may be altered by drugs through changes in the normal stimuli passing along the afferent (sensory) nerves, changes in the irritability of the vaso-motor centre itself, or by stimulation or paralysis of the efferent (sympathetic) nerves. The vaso-motor centre may be directly stimulated by drugs, indirectly through the cervical sympathetic, the vagus (when the brain is intact and the animal not narcotized), or through the general sensory nerves, when increased blood-pressure will result through contraction of the abdominal vessels. Previous section of the splanchnic nerves will prevent this rise of blood-pressure. The vaso-motor centre is also subject to irritation in changes in the respiratory gases of the blood.

The vaso-motor centre may also be inhibited, and decreased blood-pressure produced, by stimulation of the depressor nerve.

The following table, compiled by Lauder-Brunton, is introduced to facilitate reference :—

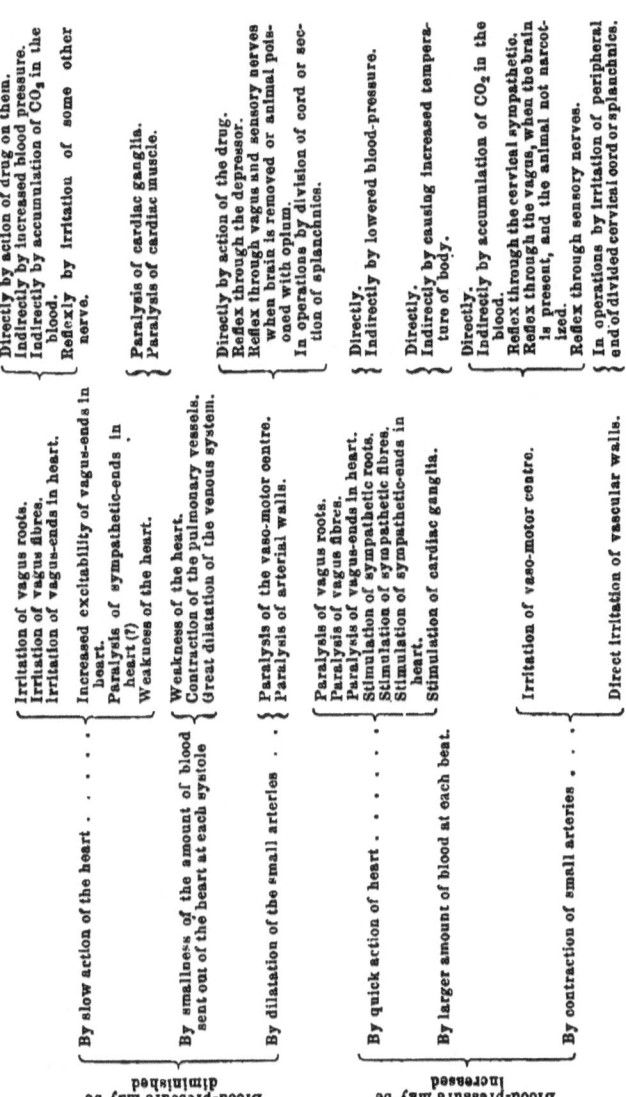

Blood-pressure may be diminished

By slow action of the heart.
- Irritation of vagus roots.
- Irritation of vagus fibres.
- Irritation of vagus-ends in heart.
  - Directly by action of drug on them.
  - Indirectly by increased blood pressure.
  - Indirectly by accumulation of CO_2 in the blood.
  - Reflexly by irritation of some other nerve.

By smallness of the amount of blood sent out of the heart at each systole.
- Increased excitability of vagus-ends in heart.
- Paralysis of sympathetic-ends in heart (?)
- Weakness of the heart.
  - Paralysis of cardiac ganglia.
  - Paralysis of cardiac muscle.
- Weakness of the heart.
- Contraction of the pulmonary vessels.
- (Great dilatation of the venous system.)

By dilatation of the small arteries.
- Paralysis of the vaso-motor centre.
- Paralysis of arterial walls.
  - Directly by action of the drug.
  - Reflex through the depressor.
  - Reflex through vagus and sensory nerves when brain is removed or animal poisoned with opium.
  - In operations by division of cord or section of splanchnics.

Blood-pressure may be increased

By quick action of heart.
- Paralysis of vagus roots.
- Paralysis of vagus fibres.
- Paralysis of vagus-ends in heart.
- Stimulation of sympathetic roots.
- Stimulation of sympathetic fibres.
- Stimulation of sympathetic-ends in heart.
  - Directly.
  - Indirectly by lowered blood-pressure.

By larger amount of blood at each beat.
- Stimulation of cardiac ganglia.
  - Directly.
  - Indirectly by causing increased temperature of body.

By contraction of small arteries.
- Irritation of vaso-motor centre.
  - Directly.
  - Indirectly by accumulation of CO_2 in the blood.
  - Reflex through the cervical sympathetic.
  - Reflex through the vagus, when the brain is present, and the animal not narcotised.
  - Reflex through sensory nerves.
- Direct irritation of vascular walls.
  - In operations by irritation of peripheral end of divided cervical cord or splanchnics.

After having determined in the manner indicated on page 113, the general action of the poison on the circulation, the comparative effects of different doses must be studied.

It must also be remembered that muscular contractions will cause an increased blood-pressure ; hence when the poison being experimented with produces convulsions, or is a respiratory poison, enough curare must be given to paralyze the motor nerves of the voluntary muscles, and artificial respiration maintained. For this purpose an ordinary bellows, run at the proper rate by a gas or water motor, or even by the hand, or any of the various forms of air-blasts, such as Sprengel's air-pump, may be used and the blast rendered intermittent by an electro-magnet by which a weight that compresses the air-tube can be alternately elevated and depressed.

In the explanation of the results obtained in blood-pressure experiments, it will not be necessary to here give a complete analysis of all the methods employed in settling each point ; enough only will be given to show the general plan to be followed.]

As already said, the causes of changes in the circulatory mechanism may lie either in an action of the poison on the muscular apparatus of the heart, its nervous apparatus, or on the bloodvessel system. The conditions which modify the functional activity of the cardiac muscle cannot be separated from those governing the other muscles of the body ; hence, general muscle poisons, especially those which exert a paralyzing influence, such as the deprivation of the blood of oxygen, will act in the same manner on the heart. But since an important part of the cardiac nervous mechanism is contained within the heart, it is often difficult to decide what effect should be attributed to action on the nervous system and what to action on the muscle. It can, however, be positively stated, that a poison acting on the muscular tissue of the heart alone, may change the force of contraction, but never produce any change in rhythm ; therefore, as a rule, it can only be held that the poison acts on the cardiac muscle when it produces progressive or total paralysis without change in rhythm; and the supposition is rendered more probable, when the drug is known to affect other muscles in a similar manner.

There is a large group of poisons, of which curare is a good example, which have no action on the muscular apparatus of the heart, but which paralyze the nerve terminations of its motor apparatus. The exact mode of termination of these nerves has not yet been determined (?), but it has been found that these poisons only act in slight degree, if at all, on the intra-muscular cardiac nerves.

The second and most usual action of a poison on the heart is on its nervous system.

[Let us suppose a case in which the drug causes quickening of the pulse; by reference to the table, we see that the heart may be caused to beat more rapidly by stimulation of the accelerator nerves or ganglia, either directly or by diminished blood-pressure, or by paralysis of the inhibitory nerves or ganglia. If the pulse is rendered quick by decreased blood-pressure, increasing the pressure by compression of the aorta, or by an injection of defibrinated blood, should slow the pulse; if, however, it should happen that the rapid pulse is associated with an increased pressure, this possibility of course does not exist. If, therefore, we assume that we are dealing with a case in which increased blood-pressure is accompanied by a rapid pulse, the question will be narrowed down as to whether the inhibitory apparatus of the heart is paralyzed, or the accelerator apparatus stimulated. If we divide both pneumogastrics in the neck before the experiment, and still find that the pulse is further increased after the administration of the drug, we can assume that the cardio-inhibitory centre in the medulla was not paralyzed, and if we find that the irritation of the central end of a divided vagus, the other being intact, can reduce the rate of pulsation, we can infer that the cause of the disturbance does not lie in the inhibitory apparatus.

Suppose, however, we find that the irritation of the central end of the vagus has no effect; then the trouble must lie either in the vagus fibres, in the heart ganglia, or in the medullary ganglion. The latter may be partially excluded by the above experiment of dividing the pneumogastrics; the question may be still more decisively answered in the following manner: It is known that after a poison has been injected into the circulation, it is only gradually distributed throughout the entire system, and its characteristic effects are only produced when the percentage of poison in the blood reaches a certain height; and of course the percentage is greatest at the point of entrance. Let us now suppose that we are dealing with a case in which either the cardiac or medullary inhibitory centre is either paralyzed or irritated; if we inject the poison into the jugular vein towards the heart and the symptoms (quick pulse for paralysis, slow pulse for irritation) instantly appear, the probability is, that the poison acts directly on the heart. But if some time, say a minute or more, is required before the effects appear, the evidence then points to implication of the centres in the medulla, and if we find that injection of the drug into the carotid causes the instant appearance of the symptoms, the evidence is conclusive.

It might, however, be necessary to determine whether the vagus trunks are paralyzed; if the cardiac ganglia are intact, this can be readily settled by testing whether their irritation slows the heart. If the cardiac ganglia should be paralyzed, the condition of the vagus may still be determined by the presence or absence of muscular contractions in the larynx after irritation of the vagus above the origin of the laryngeal nerves. After paralysis of all portions of the inhibitory apparatus are thus excluded, the conclusion can be formed that the drug acts by stimulation of the accelerator apparatus, and this can be located in the heart, if it should be·seen after section of the accelerator nerves.

The explanation of the production of a slow pulse is reached in a somewhat similar manner. Slow pulse from irritation of the inhibitory centre in the medulla, or of the vagus trunks, is excluded by section of the pneumogastrics low down in the neck, while stimulation of the peripheral ends of the vagus is rendered impossible by previously giving enough atropia to paralyze the pneumogastrics. Or the comparative degree of excitability of the vagus trunks may be tested by noting the strength of current required to slow the heart before and after the administration.

By methods similar to those here outlined, the action of a drug on the extrinsic cardiac nervous system can be determined with tolerable accuracy; the study of the action of the drug on the heart itself is, however, a matter of considerably greater difficulty.]

Since there are at least three independent cardiac centres, viz., the intrinsic cardiac centres and the two cerebro-spinal regulating centres (the accelerating and inhibitory centres), and since all these centres may be influenced by the numerous nerves with which they are connected, a poison acting on the heart through the medium of the nervous system may produce its characteristic action in several different ways; and the question is still further complicated by the fact that nearly all heart poisons act on the heart in several different ways at the same time, and the mode of action may vary in different stages of poisoning with the same drug. The following methods will serve to give some idea of the course to be followed in attempting an investigation of these points.

The action of the poison on the ganglia in the heart can be tolerably well isolated by allowing the poison to act on the excised frog's heart, or in mammals by separating the heart from the extrinsic nervous system by section of all the nerves passing to it, an extremely difficult operation,[1] or by division of both pneumogastrics and sympathetics in the neck and the spinal cord be-between the occiput and atlas. By either of these methods, however, the intracardiac ganglia are not completely isolated, since there are numerous poisons, such as

---

[1] Ludwig und Thiry, Wiener Acad. Sitzgsber. 1864, 18 Feb.

curare, nicotine, atropine, etc., which produce their characteristic effects by action on the intracardiac terminal fibres of the different cardiac nerves, especially of the vagus. In order to establish a condition of paralysis of these nerves, examinations must be made, before proceeding to the isolation of the heart, as to whether irritation of these nerves will produce the characteristic normal result after the poison has been given ; for instance, if the drug paralyzes the intracardiac endings of the pneumogastric nerve, the irritation of the nerve after the administration of the poison will fail to slow the pulse. To determine whether the slowing of the pulse or arrest of the heart, produced by the poison, is due to irritation of the vagus endings in the heart, atropia or curare, which in large doses are known to paralyze these structures, is administered before the drug, artificial respiration kept up, and the drug then given; in such a case, if the action of the poison is to slow the heart by stimulation of the vagus endings, the previous administration. of a drug, such as curare, which paralyzes these structures, will, of course, prevent the appearance of the usual symptoms.

After eliminating in this way, the possible action of the poison on the termination of the nerves in the heart, the study of the action of the drug on the isolated heart will then render it possible to form conclusions as to the action on the cardiac ganglia.

[From reasons already given, the heart of the frog is much better suited for the study of drugs than is that of the mammal, though recent improvements in the methods of research have rendered the heart of warm-blooded animals much more accessible for this purpose. The methods of studying the local action of poisons on the heart *in situ* have been already given; for the excised heart, several plans may be followed. The old modes of study, alluded to on page 88, have now been universally supplanted by the methods of investigation introduced by Ludwig; his plan was to keep the heart supplied with serum and attached to a manometer, by which

the pulsations of the heart could be recorded. His original instrument has been considerably modified by several
investigators, and several new forms of instruments for
this purpose are now in use; of these, only two forms
will be described, accounts of the others can be found in
the various physiological hand-books.

The apparatus used by Ludwig and Coats in their experiments on the vagus nerve is shown in Fig. 27 ; it is

Fig. 27.

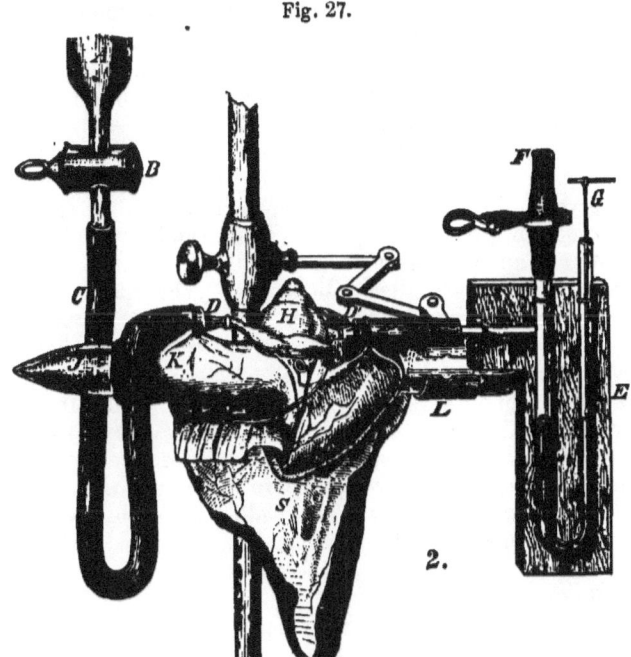

Ludwig and Coats's frog manometer.

the simplest form of frog manometer, and can be readily
extemporized. It consists of a manometer E, connected
by a canula D' with the bulbus aortæ of a frog; at D
is another canula inserted into the sinus venosus and con-

nected by the tube C with a reservoir A containing diluted rabbit serum, or even normal salt solution. J is a heavy glass rod, moving on a sliding clamp, for holding the frog. The frog's heart is prepared by destroying the brain and spinal cord and then cutting across the body below the liver so as to remove the lower extremities; the sternum and forelegs are then removed, leaving only a flap of skin large enough to cover the heart, which is exposed in the usual manner; one canula is then inserted in one aorta, pushed into the bulb and bound fast, while the other aorta is ligated; another canula is then inserted in the sinus venosus. The liver and lungs are then removed and an opening made in the stomach, and the glass rod J passed through the mouth and down the œsophagus; the aortic canula is then connected with the manometer and the venous canula with the reservoir, the stop-cock B opened and the serum allowed to flow through the heart and out at F, until all air bubbles are displaced and the heart and vessels completely filled with serum. The clamp on F is then closed and the pressure bottle raised to such a height that there is a certain tension exerted on the heart even during diastole. This method of using this appa-. ratus, in which there is no circulation, the serum simply being forced out of the ventricle at each systole and falling back at each diastole, is especially suited for the study of drugs on the vagus nerve. The vagus is seen in the drawing, and may be readily found below the greater horn of the hyoid bone lying alongside of the laryngeal nerve, which can be readily recognized by tracing it to its destination.

After normal tracings have been taken with this apparatus and the effects of the vagus tested, some of the poison may then be added to the serum in the reservoir and the results noted. In many cases, it is better to have an active circulation through the heart, and to be able to substitute normal for poisoned serum. This can be readily accomplished by having two reservoirs standing on the same level, the one containing normal

ACTION ON THE CIRCULATORY APPARATUS. 119

serum, and the other serum containing poison; their flow
through the heart can then be regulated by a two-way
cock, while by opening slightly the clamp *F*, the serum
that is pumped through the heart can be allowed to
escape. In the comparison of results obtained in this
manner with normal and poisoned serum, care must be
taken that the conditions are always uniform; that the
resistance at *F*, and the pressure in the reservoirs are
always the same. By varying the resistance at *F*, the
effects of increased or diminished capillary resistance

Fig. 28.

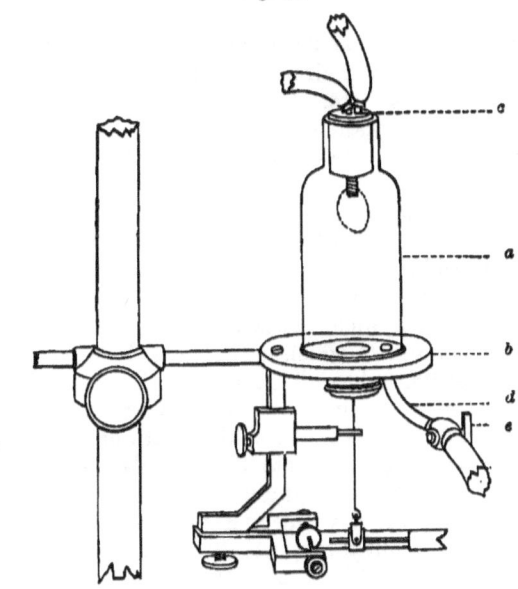

Roy's heart apparatus.

may be imitated and the effect of varying blood-pressure
studied on the action of the heart.

By means of an instrument devised by Dr. C. S. Roy[1]

[1] Journ. of Physiol., vol. i. No. 6.

the accurate study of poisons on isolated portions of the
frog's heart is greatly facilitated. The instrument is
represented in Fig. 28. The small bell-glass (*a*) rests
on a round plate of brass (*b*) to which it is fixed by the
aid of a little stiff grease. In the upper opening of this
vessel is fitted a short glass canula, which is perforated
to allow the passage of the heart canula; inside this
canula is a second tube of metal measuring about 1 mm.
in diameter. It extends from the lower extremity of the
canula to a point about 5 mm. from its upper end, where
it passes through its canal, and projects for a sufficient
distance to allow of an India-rubber tube being tied on
it (*c*).

By means of this canula, diluted blood, or other fluid,
can be kept constantly circulating through the auricle or
ventricle, which is fastened on it, the rapidity of the flow
being regulated by the difference in height of the two
reservoirs which are in connection with the two tubes of
the canula.

In the brass plate on which the bell-jar rests are two
openings, one of which forms the inlet to a short tube *d*
which is provided with a stop-cock *e*; the other opening
is situated in the centre of the brass plate and forms the
inlet to a short cylinder closed below by a non-elastic
flexible membrane to which is attached, by a piston-like
disk and needle, a long light lever. The ventricle, or
auricles, as the case may be, having been fastened on,
and the canula and reservoirs filled with diluted blood,
the heart is introduced into the bell-glass (which has
been previously fixed on the plate), and its cavity filled
with olive oil. On now opening the stop-cock (*e*) the
oil begins to flow out through the tube *d*, and renders the
pressure within the bell-glass sub-atmospheric; when
the piston has thus been drawn up to the point represented in the figure the stop-cock is closed. With the
apparatus thus arranged, each contraction of the heart
will cause an elevation of the lever, and each relaxation
a fall, from the varying volume of the contracting heart.
Observations on the isolated ventricle may be made by

cutting away the base of the heart nearly down to the auriculo-ventricular sulcus, inserting the canula into the ventricle through one of the auricles and then binding it fast by a ligature passing around the auricular wall near the ariculo-ventrical groove. Where the ventricle is very small the auricular septum may interfere with the introduction of the canula; in such cases the auricular septum should be slit through with a pair of very fine blunt-pointed scissors.

To fasten on the auricle to the perfusion canula the lower two-thirds of the ventricle is clipped off with scissors and the auricular septum slit through with blunt-pointed scissors, one blade entering each auricle from the ventricular aspect, great care being of course taken to avoid cutting the walls of the auricles. The venæ cavæ superiores and inferior are then ligated, the position of the ligature around the sinus venosus varying in different experiments; the end of the canula is then in-troduced into the auricular cavity from the opened ventricle, and is fastened by a ligature passing around the sulcus, the upper third of the ventricle or around the lower part of the auricular walls as the case may require. The movements of the sinus venosus may also be studied in the same manner by introducing the canula into its cavity. It is seen that by this method it is rendered perfectly feasible to study the action of poisons on each separate portion of the heart; tracings may first be taken with the organ supplied with diluted blood and then with blood containing definite proportions of the drug. It is possible to experiment on the toxic changes which the heart undergoes as regards its electric irritability by using the canula as one electrode and by surrounding the canula where it passes through the stopper with rubber, and outside of it a sheet of tin-foil whose projecting edge is cut into a fringe; outside, the tin-foil is connected with the other pole of the induction apparatus, while inside any portion of the fringes may be placed in contact with the heart and serve as the second electrode.

The recent application to the mammalian heart by
11

Prof. Martin,[1] of Baltimore, of the Leipzig method of maintaining circulation through the organs of warm-blooded animals has rendered possible the study of drugs on the isolated mammalian heart. The principle of this method is to prevent circulation through all parts of the body of a warm-blooded animal but the heart and lungs; from want of blood, the brain, spinal cord, and sympathetic ganglia soon die, and so the heart is liberated from the control of nerve centres outside of itself.

The animal being tracheotomized and narcotized, the carotids are exposed and tied, and canulæ placed in their central ends; the vagi are then divided in the neck. The next step is to expose the heart and great vessels by resecting the front and sides of the thorax, all bleeding vessels being ligated. The right and left subclavian arteries are then tied below the origin of their first branches, thus cutting off nearly all blood from the head. Next a metal canula, curved at one end so as to present a long limb and a short limb at right angles to one another, is inserted into the aorta just above the diaphragm and pushed up until its end reaches the arch, where it is bound fast, thus blocking all circulation through the systemic arteries, with the exception of the coronaries. The next step is to tie the systemic veins leading to the right auricle; a ligature is placed around the inferior vena cava above the diaphragm, another around the vena azygos near its entry into the superior vena cava, and the latter is then ligated on the cardiac side of its last tributary. On the cardiac side of this ligature a large tube, in communication with a flask containing defibrinated diluted blood, is introduced, the carotids opened, and all the blood previously present in the heart and lungs displaced by defibrinated blood. A thermometer being inserted in the left carotid, and the right connected with the manometer tube, the animal is then transferred to a warm moist chamber.

[1] Trans. of the Med. and Chir. Fac. of Maryland, April, 1882, p. 203.

The aortic canula is connected with a long rubber tube having at its distal end a bent glass tube from which the blood, forced out by each contraction of the left ventricle, is poured into a funnel; from this funnel a tube leads to a Mariotte's flask exactly like that in connection with the right auricle. The blood taken by the right heart under definite pressure from one Mariotte's flask is thus pumped into another, from which, by changing a couple of stop-cocks, it can a second time flow into the right heart. Varying blood-pressure can be produced by elevating or depressing the end of the aortic exit-tube, while the addition of the poison to the blood in the venous reservoir will enable its action on the isolated heart to be studied.

In the attempt to localize the action of heart poisons on different portions of the ganglionic apparatus of the heart, one of two ways may be followed, though neither in the present state of cardiac physiology deserves to be designated as a method. Either the poison may be administered to pulsating fragments of the frog's heart by means of Roy's apparatus, and some conclusion attempted from the known anatomical peculiarities of the part, or the method of antagonisms may be followed. In general, the latter, although it is true that its data are largely based on assumptions, will lead to the most reliable results. Thus in the diagram, Fig. 29, the fibres represented by the dotted line A, are said to be paralyzed by nicotine, the ganglion I irritated by muscarine, and the fibres B paralyzed by atropine. Suppose, therefore, we find that a drug produces increased rapidity of the pulse in an excised or isolated heart; we

Fig. 29.

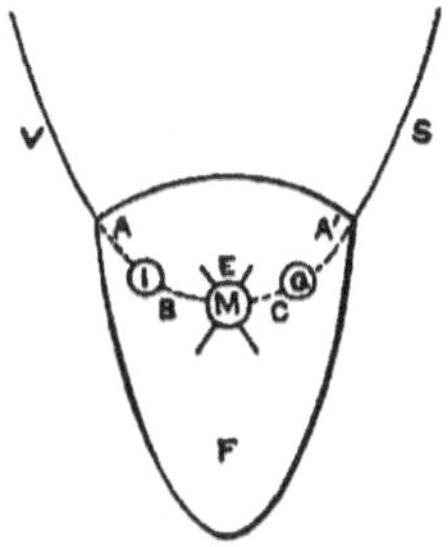

Diagram of the hypothetical nervous apparatus of the heart.

may first irritate the vagus, we find it fails to slow the pulse; we then irritate the sinus venosus, still without effect; we then add a few drops of muscarine and still find that the heart is not slowed. We then know that the fibres B must be paralyzed by the drug in question. Or suppose a case in which muscarine could slow the pulse, while irritation of the vagus failed; we then know that the fibres A were paralyzed. If the drug should slow the pulse, and the subsequent administration of nicotine should not restore the normal rate, we would suppose that some portion of the ganglionic apparatus nearer the motor centre than A must be affected, and if we found that atropine would remove the effect of the drug, it would be probable that the drug in question produced slowing of the pulse by stimulation of the inhibitory ganglion I.

As regards the action on the accelerator apparatus our knowledge is not so complete. When we find that a drug quickens the excised heart, we have one of two possibilities to consider; either the paralysis of the inhibitory apparatus or the stimulation of the accelerator apparatus. The former may be excluded by previous paralysis of the inhibitory ganglia by atropia. If now the drug produces quickening we know that it must be by action on the accelerator apparatus; further than this we cannot at present go, as the list of drugs which act on the accelerator ganglia is very limited and not yet well worked out.]

The third cause of modifications in the action of the heart and in blood-pressure produced by a poison, lies in the condition of the peripheral vascular system. The degree of contraction of the bloodvessels, particularly of the arteries, not only influences the degree of blood-pressure in the vessels, but also the frequence and energy of the heart's contractions. Thus, by ligation of any large arterial trunk, such as the descending aorta, the pressure in all the other arteries and in the left side of the heart can be so increased that the distended heart will be only able to perform very feeble contractions. It has also been experimentally determined that the calibre of the smaller arteries is subject to variation depending upon the degree of contraction of their muscular walls, and that this contractility is governed by the impulses coming, by the efferent vaso-motor nerves, from the principal vaso-motor centre in the medulla [1]

Many poisons influence the degree of arterial contraction either by direct action on the arteries (either by action on their muscular fibres or on the hypothetical peripheral vaso-motor ganglia), or on the vaso-motor centre in the medulla, so as to produce either paralysis and dilation of the arteries with a consequent reduced blood-pressure, or a reduction in calibre with a consequent great increase in blood-tension. Experiment has further shown that each increase in pressure is usually accompanied by a reduction in the pulse, and each reduction in pressure by an increase in the pulse.[2]

It is, therefore, unwarrantable to form any conclusion as to the cause of modifications in the heart's action until the possible reflected influence of changes in the conditions of the bloodvessels has been considered. The reflex influence of the central vaso-motor centre on the heart can be eliminated, without interfering with the activity of the accelerating nerves, by section of the spinal cord on the level of the second dorsal[3] vertebra, or by division of the splanchnic nerves.[4] But even with this procedure the possibility of direct toxic action on the muscular fibres of the arteries still remains. Observations as to the condition of the bloodvessels are most readily made on the ear of the rabbit, especially after depilation with sulphide of calcium,[5] in the wing of the bat, or in the retinal vessels of all animals capable of being examined with the ophthalmoscope; the mesentery or swimming bladder of the frog can also be used for the same purpose.

[1] For the exact location of this centre see Owsjannikow, Säch. Acad. Ber. 1871, p. 135.
[2] See Ludwig and Thiry, Wiener Acad. Sitzgsber., 1864, Feb. 18.
[3] V. Bezold, Untersuch. aus d. physiol. Lab. in Würzburg, 2 Heft, 1867.
[4] See M. and E. Cyon, Arch. f. Anat. u. Physiol., 1867, p. 395.
[5] See Samuel, in Moleschott's Untersuch., ix. p. 654.

[Exclusive, then, of modifications dependent directly upon the heart, the blood-pressure may be modified by the direct action of the drug on the afferent vaso-motor nerves, on the vaso-motor centre in the medulla (and cord?) and on the efferent nerves. Consequently, when it is found that a drug produces a reduction in blood-pressure, after the exclusion of the causes depending on cardiac action, the condition may be due to paralysis of the vaso-motor centre from direct action of the drug, to paralysis of the afferent or efferent vaso-motor nerves, to irritation of the depressor nerve, or to direct local action on the bloodvessels.

When the cause has been located in the vaso-motor apparatus, the precise seat of the paralysis can only be determined by working from the periphery to the centre; thus the normal, or abnormal, condition of the arterial walls must be first determined, then that of the efferent vaso-motor nerves, then the vaso-motor centre, and finally that of the afferent vaso-motor nerves. In most cases it is extremely difficult to separate direct toxic action on the bloodvessels from action on the efferent vaso-motor nerves, though some deductions may be drawn from the characters of the circulation in excised organs; the methods for carrying on these studies will be given under their appropriate heads.

If the poison produces reduced blood-pressure from direct action on the vascular walls, whether on their nerve-ending or muscular fibres, we would expect that after arterial tension has been reduced through section of the cord, and the influence of the vaso-motor centre thus eliminated, the administration of the drug would be followed by a still more marked fall in pressure.

Local action on the bloodvessels may be excluded, as was done by Filehne in the case of nitrite of amyl, by maintaining artificial circulation with normal blood through the vessels of the external ear of a rabbit, and then administering the poison either by injection into the venous system at large, or through the trachea when in the form of a vapor. Should the vessels then dilate,

local action on their walls or on the nerve endings would
be excluded.

Another method, also employed by Filehne for the
same purpose, is to maintain a condition of moderate
contraction of the auricular vessels by stimulation of the
cervical sympathetic on one side with a weak interrupted
current; if dilatation should not appear on that side after
administration of the drug, but exist on the ear of the
opposite side, the dilatation could be attributed to diminished tonus of the vaso-motor centre.

The irritability of the efferent vaso-motor nerves may
be determined by irritating the dorsal spinal cord, or the
splanchnic nerves, when, if the efferent vaso-motor fibres
preserve their functions, the blood-pressure will be greatly
increased from contraction of the abdominal arterioles;
should they or the arterial walls be paralyzed, no such
rise will be produced. Or the central end of the divided
cervical sympathetic may be stimulated in a rabbit and
the auricular vessels directly inspected; should they contract, it will be evident that the vaso-motor paralysis is
located in the centre or in the afferent nerves.

The irritability of the vaso-motor centre may be determined by screwing one gimblet electrode into the
occipital bone and the other into the atlas, until their
points penetrate the neural cavity, and passing an induced current through them. Or the vaso-motor centre
may be irritated by compressing the carotid arteries in
the neck by raising them on threads; in a normal condition this experiment produces an increase in blood-
pressure. If the blood-pressure is increased by either
of these modes of stimulation, it may be considered demonstrated that the vaso-motor centre and efferent nerves
preserve their functions, and it will then be necessary
to determine the condition as regards the power of transmitting impression possessed by the afferent vaso-motor
nerves. This is accomplished by irritating the central
end of the divided sciatic nerve, a procedure which
normally is followed by an increase in blood-pressure.

Should all these experiments demonstrate a normal

state of irritability of the vaso-motor apparatus, attention must then be directed to the depressor nerve. This
nerve, which is a branch of the pneumogastric nerve, or
rather a root of the latter, which in the rabbit joins it at
the level of the superior laryngeal nerves, possesses the
power through its irritation of inhibiting the vaso-motor
centre in the medulla and thus producing a marked fall
in blood-pressure. If, therefore, both these nerves are
cut in the rabbit before the administration of the poison
the possibility of their influence in the production of
reduced blood-pressure will be excluded.

It should, moreover, be always remembered that drugs
which produce paralysis of the vaso-motor system usually,
especially with small doses, first cause a condition of
irritation of this apparatus, hence the fall of blood-pressure is generally preceded by an initial rise.]

INDIRECT RESULTS OF CIRCULATORY DISTURBANCES.—
In cold-blooded animals marked disorders of the circulatory apparatus are without effect on other functional
activities ; it is only when they are long continued that
general disturbances appear, and after a time, after complete arrest of the heart, the animal gradually becomes
more and more sluggish, its loss of power gradually passing into complete paralysis and death.[1] The secondary
cause of these disturbances, after arrest of the circulation, probably lies in increasing deprivation of oxygen
affecting the central nervous system and muscles simultaneously.   Increased rapidity of heart pulsation is
entirely without effect.

In warm-blooded animals, on account of their constant
need of fresh supplies of oxygen, every considerable reduction in the rate of the heart's beats, and especially
arrest of the heart, is accompanied immediately by the
gravest general disturbances, and the recognition of this
interdependence of general functional activities and the
state of the circulation, first pointed out by Rosenthal,[2]

[1] [See in this connection Ringer and Murrell, Journ. of Physiol.,
vol. i. No. 1. p. 72.]
[2] Arch. f. Anat. u. Physiol., 1865, 601.

is to be regarded as one of the most important advances in scientific pharmacology.

Arrest of the heart causes a complete stagnation of the blood in all the vessels, and as a consequence we have on the one side, a cessation of absorption of oxygen from arrested pulmonary circulation, and on the other side, the different organs are supplied with a diminished quantity of blood which rapidly gives up its oxygen and becomes loaded with carbonic acid. This interruption in the oxygenation of the blood in warm-blooded animals rapidly destroys the functions of all organs and soon leads to general systemic death. Before, however, death occurs, there appears a series of phenomena depending upon the arrested circulation in the medulla oblongata. At first the respiratory centre is abnormally stimulated by the altered character of the blood, and when the *venosity* of the blood passes a certain degree, the stimulation extends to the neighboring motor and vaso-motor centres in the medulla, and contraction of all the small arteries, and then general convulsions follow. Hence, arrest of the heart is followed by the same train of symptoms as interruption of the circulation in the brain by ligation of the cerebral arteries or veins (which produced the same arrest of cerebral circulation as stoppage of the heart), or as interference with respiration; namely, in the first place, increasing vigor of respiration up to dyspnœa, then general convulsions and arterial spasm, while the phenomena of asphyxia first make their appearance when the amount of oxygen in the blood of the medulla oblongata has fallen so low that the nerve centres lose their irritability. These phenomena will be more closely studied under the respiratory changes produced by poisons.

It consequently follows from what has been said that arrest of the heart, or even every considerable reduction in the heart's activity, must in warm-blooded animals cause dyspnœa, general convulsions and asphyxia. When therefore a poison causes convulsions in warm-blooded animals and not in frogs, it must always be determined whether the drug does not in the first place cause stoppage of the heart.

In this way other nerve centres, especially those governing the movements of the intestine,[1] and contractions of the uterus,[2] are influenced like the respiratory and vaso-motor centres by increased venosity of the blood, and consequently increased intestinal peristalsis and uterine contractions may be results of interference with the circulation.

The consequences of less grave circulatory disturbance are to be explained as depending upon alterations of blood-pressure (see p. 91). Nearly all functions are intimately dependent upon the degree of blood-tension in the organ with which they are associated, and therefore every marked change from the normal blood-pressure, through toxic action on the heart or bloodvessel system, soon leads to functional disorders.[3] The most marked and the earliest of these disorders occur in the functions of the cerebrum, where diminished blood-pressure leads to dizziness and syncope, and increased pressure to sense disturbances, delirium and loss of consciousness.

Section II.—Action on the Respiratory Apparatus.

A general idea as to the condition of the respiratory apparatus may be gained by mere inspections of the thorax. A greater degree of accuracy is attainable when inspection of the diaphragm is rendered possible by opening the abdomen ; or, without opening the abdominal cavity, by the insertion of a long needle through the body walls into the diaphragm. In order to reproduce

[1] Mayer u. von Basch, Wiener Med. Jahrbücher, 1872.
[2] Oser u. Schlesinger, Weiner Med. Jahrbücher, 1872.
[3] The best statement of these facts is to be found in a lecture by Ludwig, Die Physiologischen Leistungen des Blutdruckes, Leipzig, 1865.

the respiratory movements graphically Rosenthal's phrenograph or Marey's pneumograph may be used.

In certain cases it is necessary to determine the volume of the respiratory movements; for this purpose Hutchinson's spirometer or an ordinary gas-meter, connected with the trachea, may be employed.

[The changes in the frequency and rhythm of the respiratory movements may be graphically represented by means of Marey's tambour, either arranged as in the figure (Fig. 30) or by connecting the tube *d* directly with the tracheal canula, which must then be provided with an opening at one side to enable the animal to obtain fresh air.

Knoll[1] recommends the insertion of the animal in a box which can be closed air-tight, the trachea of the animal being connected with the exterior by a tube passing through the top of the box, while the respiratory movements are recorded by connecting the interior of the box by a tube with a Marey's polygraph.]

a. DYSPNŒA.—The most frequently observed effect of poisons on the respiratory apparatus is dyspnœa, made evident by an increased vigor of the respiratory movements and action of the accessory muscles of respiration; usually, the frequency of the respiratory movements is also diminished.

The cause of dyspnœa is invariably an irritation of the respiratory centre in the medulla oblongata; this irritation is always to be found in the blood which acts as an excitant to the respiratory centre in proportion as it becomes poorer in oxygen and richer in carbon dioxide.[2] These morbid states of the blood, which are therefore the final cause of dyspnœa, can exist either in the bloodvessels of the medulla or the head, as has already been

---

[1] Sitzber. der Akad. zu Wein., 3 Abth. Bd. lxviii. s. 245.

[2] These alterations of the blood are mutually interdependent; the question as to which of the two conditions is the true excitant still remains undecided.

Fig. 30.

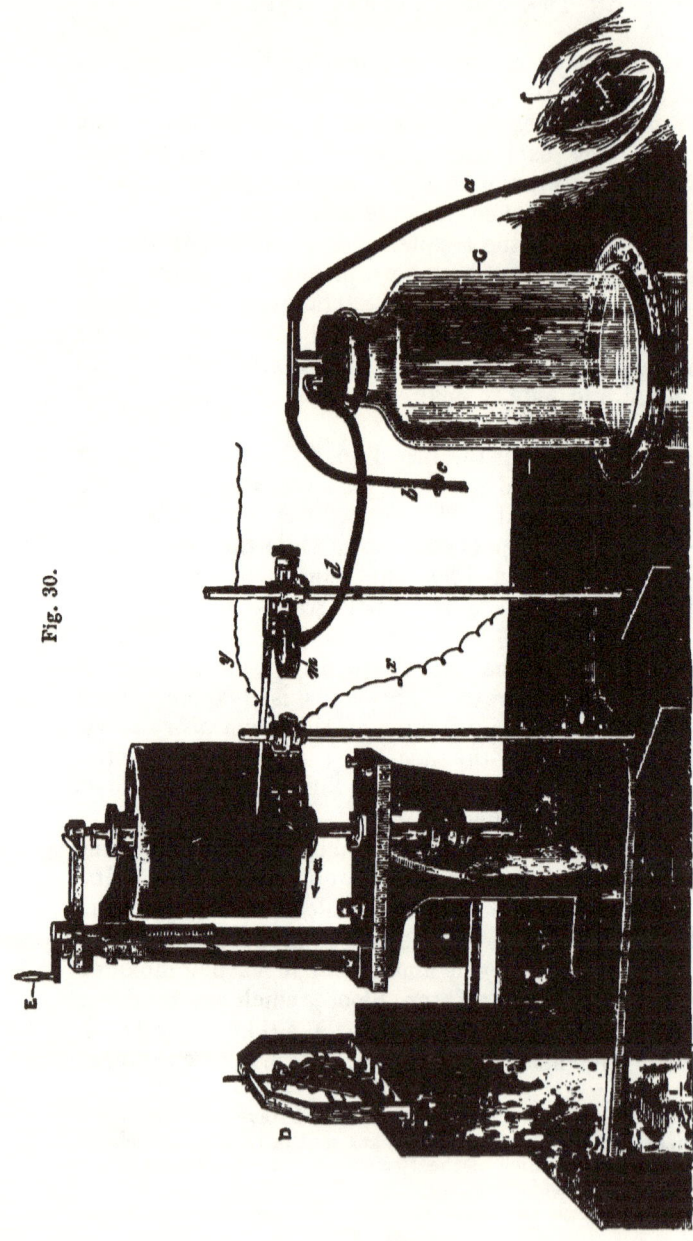

Apparatus for taking tracings of the movements of the column of air in respiration. (From Foster's Physiology.) The tracheal canula *t* is connected by the tube *a*, with the large jar G, the continuation of the tube, *b*, being capable of being partially or totally closed by the clamp *c*. The interior of the jar is connected, through the tube *d*, with the tambour *m*, whose lever, *e*, writes on the revolving drum A. Below the lever is seen a time-marker in connection with an electro-magnet, the current through whose wires *z* and *y* is broken by a metronome or similar apparatus.

mentioned on p. 128, or they may prevail in the blood at large ; this latter condition is the more usual.

The reduction in percentage of oxygen in the blood and the excess of carbon dioxide, or, as Hering terms it, the increased *venosity* of the blood, may be due to any or all of the following conditions: 1. Diminished or retarded absorption of oxygen in respiration. 2. Diminished exhalation of carbon dioxide in respiration. 3. Expulsion or abnormal consumption of the oxygen originally in the blood. 4. Abnormal absorption of carbon dioxide.

The first two of these conditions are ordinarily produced by interference with the respiratory movements, and may be produced by poisons through paralysis of the respiratory apparatus, either of the respiratory centre, nerves or muscles. In this case, however, to produce *dyspnœa*, the paralysis must be confined to individual respiratory muscles ; or they may be due to interference with the pulmonary circulation (see heart paralysis, p. 129), or finally, to inability for absorption of oxygen by the blood. This latter cause, with the exception of dyspnœa produced by heart-poisoning, is the most frequent factor in the toxic production of respiratory difficulties. The blood can be unfitted for absorbing oxygen either through alteration of the hæmoglobin, or through destruction of the red corpuscles.

The third of the above-mentioned conditions which may act as causes of dyspnœa, the expulsion of the oxygen from the blood, can naturally only exist as such when it is produced so rapidly that the lost oxygen cannot be replaced with sufficient rapidity, to preserve a normal proportion, by the fresh oxygen taken in respiration, or when the blood is at the same time and by means of the same agent unfitted for absorption. Sulphuretted hydrogen is an example of a poison which produces the first of these conditions, carbon mon-oxide, of the second.

The fourth condition may occur when an atmosphere abnormally contaminated with carbon di-oxide is inspired.

If, now, this poverty of oxygen and richness in carbon

12

di-oxide of the blood continues, the dyspnœa passes into
general convulsions as the increasing irritation extends
to other centres in the medulla.   Should the deprivation
of oxygen still advance, the irritability of both the nerve
centres and the muscles gradually fails, the convulsions
cease, and are replaced by asphyxia, which is not death
so long as the heart continues to beat and still possesses
the power, through introduction of oxygen into the blood,
of bringing the organism back to its normal condition.
Should this occur, as, for example, through artificial respiration, the irritability of the nerve centres is first re-
newed, and convulsions are again produced ; the rapid
disappearance of the irritation soon causes the convulsions
to be replaced by dyspnœa and this finally gives place to
normal respiration.   The consequences of dyspnœa cannot be specified, since dyspnœa is essentially a compen-
satory process, tending to remove the abnormal condition
of the blood through increased absorption of oxygen and
exhalation of carbon di-oxide by deeper breathing.

b. CESSATION OF RESPIRATORY MOVEMENTS.—Besides
dyspnœa, poisons may produce weaker respiratory movements or cause their complete suppression.   The causes
of these conditions may, in general, be either:—
1. REDUCTION OF THE RESPIRATORY STIMULUS THROUGH
SATURATION OF THE BLOOD WITH OXYGEN AND DIMINU-
TION IN ITS CARBON DI-OXIDE.—This condition, which
can scarcely be regarded as toxicological, is termed *apnœa*,
and may be produced by vigorous artificial respiration ;
it is made use of experimentally when it is desired to
study successive stages of the general action of a drug
whose administration is otherwise followed by dyspnœa.
2. REDUCTION IN IRRITABILITY OF THE RESPIRATORY
CENTRE.—This condition may be the result of either
a direct action of the poison, or the consequence of such
a reduction in the amount of oxygen in the blood that the
nerve centres are no longer irritable, a condition which
always occurs in the last stage of dyspnœa (see asphyxia,
above).   It is directly productive of death, since not only

the respiratory centre, but all the other centres, especially those of the heart, become simultaneously paralyzed either from the same cause or from the cessation of respiration.   If the asphyxia is the result of a direct action of the poison on the respiratory centre, the activity of the heart, and therefore life, can be preserved by artificial respiration.   If, on the other hand, the asphyxia is due to an absence of oxygen in the blood, artificial respiration will only prove effective when mere increased access of oxygen will serve to supply the deficiency.

[Respiratory changes may be referred to direct action of the drug on the respiratory nerve centre when they occur after section of the vagi, and after the influence of circulatory changes has been excluded.   This point may be determined by injecting the drug into the carotid artery toward the brain, while a blood-pressure observation is made in the crural artery ; if respiratory changes occur before any disturbance of the circulation, the latter may be excluded as the active cause.   Of course the possibility of the effects of the drug being due to alterations in the blood or its gases must be taken into consideration.]

3. PÁRALYSIS OF THE RESPIRATORY MUSCLES.—This condition will be produced by drugs producing general paralysis, e. g., curare.   Death results from deficiency in absorption of oxygen, ordinarily without preliminary dyspnœa or convulsions ; in such circumstances life may be preserved by artificial respiration.

c. ALTERATIONS IN THE FREQUENCY OF THE RESPIRATORY MOVEMENTS —The rate of respiratory movement is dependent upon the rhythmical functions of the respiratory centre, upon the condition of excitation of the regulating nerves, especially those running in the pneumogastric, and upon mental or cerebral stimuli.   Drugs may change the rate of respiration through any one of these paths: ordinarily the *modus operandi* may be accurately enough determined.

Cerebral or mental sources of stimuli on the respiratory centre may readily be established in man, and in

animals they may be excluded through previous narcotization or extirpation of the cerebral cortex.

Irritation of the regulator nerves, at least when starting from their peripheral terminations, may be eliminated
by section of the nerve trunks.

Changes in frequency persisting after exclusion of the
two preceding modes of action must depend upon the
direct action of the poison on the respiratory centre.

Changes in the rate of respiration must be very
marked to produce any evident general results. It
should, however, be remembered that slowing of the respiration may be an introductory symptom of complete
cessation of respiration, increased respiration, of tetanus.

In both cases the consequences will be the same, in a
general way, as those which follow complete cessation
of respiration.

d. APPEARANCES IN THE LARYNX.—Poisons may affect
either the sensory or motor functions' of the larynx.
Insensibility of the larynx, ordinarily only one sign of
more general anæsthesia, interferes with the normal pro-
tective influence exerted by the larynx over the lungs
by the absence of the possibility of reflex closure of the
glottis ; a similar danger may also be produced in paralysis of either the laryngeal nerves or muscles.

On the other hand, poisons may cause spasm of the
glottis, and so interfere with normal respiration, either
reflexly by irritation of the sensory nerves, as in inhala-
tion of irritating gases and vapors, or by direct action on
the nerve centres or muscles.

SECTION III.—**Action on the Digestive Apparatus.**

Under this head, which has been less studied, and is,
therefore, more obscure than any other branch of pharmacology, distinction must be made between changes in
the movements of the digestive organs, the production of

abnormal sensations, alterations in the secretions, and, finally, changes in the digestive processes.

a. ALTERATIONS IN THE MOVEMENTS OF THE DIGESTIVE ORGANS.—1. MOVEMENTS OF THE JAWS.   The only toxic effect evidenced by movements of the jaws [except the masticatory movements which follow the introduction of drugs by the mouth, or their excretion, when perceptible to the taste, by the saliva] is trismus, a tetanic spasm of the muscles of mastication which is generally an introductory symptom of general convulsions (see Nervous System), and which, with the exception of the prevention of the prehension of food, leads to no special consequences worthy of separate study.

2. DEGLUTITION.—Deglutition can be hindered by the action of poisons, either through action on the motor apparatus, when it is merely a symptom of general paralysis (see Nervous System), or by alteration of the secretions of the mouth and pharynx, as, for example, the dysphagia produced by belladonna poisoning.   Whether spasm of the muscles of deglutition, as occurs in hydrophobia, may be produced by the action of poisons, is unknown.

3. MOVEMENTS OF THE STOMACH.—As the knowledge which we possess as to the conditions modifying the physiological movements of the stomach is of the most limited character, it follows that nothing can be said as to the effects of drugs on this function of the digestive organs.   As regards the production of vomiting by toxic action, our knowledge is a little more complete.

Vomiting is a complicated co-ordination of various muscles, having as a result the emptying of the contents of the stomach into the pharynx.   With the exception of the opening of the cardiac orifice,[1] the role of the stomach in vomiting is purely passive, the act being largely due to rhythmic contractions of the diaphragm and abdominal muscles.   The co-ordination of these muscles is governed by a nerve centre lying in the medulla oblongata or

---

[1] Schiff, Moleschott's Unters., x. 353.

brain[1] which is capable of being set into activity by
stimuli directly brought to it by the blood, or reflexly
through various centripetal nerves, especially those coming from the digestive apparatus. The determination as
to which of these modes is concerned in the production
of vomiting by drugs cannot be reached by merely varying the mode and location of administration of the drug,
since the production of emesis after venous or hypodermic injections does not prove a direct action on the
centre; for the substances are carried by absorption to
the stomach, and may there serve as reflex stimuli.
This statement is proved by the fact that tartar emetic
requires larger doses and a longer time to produce
emesis when injected into a vein than when given by
the stomach, and even in the former case the presence
of antimony can always be detected in the vomited
matters.[2] It is, therefore, still doubtful whether the
centre is capable of direct stimulation.

Different animals vomit with different degrees of readiness; while birds, dogs, and mice, vomit with the greatest
ease, the contrary is the case with rabbits and frogs.

Although, as has been said above, the stomach is passive in the act of vomiting, it is conceivable that drugs
may at the same time produce active contractions of the
stomach; to verify this, the voluntary muscles should be
paralyzed with moderate doses of curare and artificial
respiration kept up, when the conditions of the stomach
may be directly inspected.

The consequence of the act of vomiting is the removal
of the contents of the stomach, and, therefore, the partial
or total removal of the poison; if it is desired to study
the general action of a poison which produces vomiting,
the drug should first be administered by some other
means, as hypodermically, and if vomiting still occurs, it
must be prevented by curare,[3] or by ligature of the œso-

[1] Hermann u. Grimm, Arch. f. d. Ges. Physiol., iv. 205.
[2] Hermann, Arch. f. d. Ges. Phys., v. 280.
[3] Giannuzzi, Centbl. f. d. med. Wissen., 1865, i.

phagus, a procedure often employed in the older experiments.[1] [In many cases vomiting may be prevented by section of the vagi, since these are the nerves by which the afferent impulses which cause the relaxation of the cardiac sphincter reach the medulla.]

4. MOVEMENTS OF THE INTESTINES.—Alterations in the movements of the intestines, such as increase, diminution, or suppression of the peristaltic motions, cannot be clearly studied either by inspection or palpation without opening the abdomen, and it is therefore generally advisable to expose the abdominal contents by an incision in the linea alba ; but since the rapid loss of heat and drying of the intestines lead to changes in their circulation, and consequently to changes from the normal motions which might erroneously be attributed to the action of the drugs, it has been recommended with some show of success to immerse the animal, before opening the abdomen, in a bath of salt solution, $\frac{1}{2}$ per cent., heated to the body temperature, whereby all cooling and access of air is prevented; under such circumstances artificial respiration must be kept up through a tracheal canula and rubber tube.[2]

[Salvioli[3] employed the following method for studying the movement of the small intestine. A piece of jejunum is excised with its mesentery, from a rabbit, laid on the inner surface of a piece of excised abdominal wall in a warm, moist chamber, and a mixture of 30 parts calvesblood and 70 parts $\frac{3}{4}$ per cent. salt solution, well shaken up in the air, conducted through its bloodvessels ; one or more light levers resting on the surface of the intestine serve to register its movements. The action of drugs on the peristaltic movements may be studied by adding the poison to the circulating fluid ; thus Salvioli found that nicotine caused violent intestinal contractions and narrowing of the bloodvessels, while opium and atropine pro-

[1] See Orfila's Toxicologie, 1839, 1. 29.
[2] Sanders-Ezn u. van Braam Houckgeest, Pflüger's Arch. vi. 266.
[3] Arch f. Anat. u. Phys., 1880, s. 95.

duced the reverse. For particulars as to this method, as well as for the relations observed between the blood-pressure and the peristalsis, reference must be made to the original memoir.]

Departure from the normal degree of peristaltic motion may be due either to direct action on the muscles of the intestine or on its ganglia, on the extrinsic motor or inhibitory (splanchnic) centres, or indirectly to respiratory or circulatory changes.

· The experiments necessary for the proof, by exclusion, as to which of these modes of action is concerned, such as irritation and section of the appropriate nerves, will readily suggest themselves.

[Direct action of a drug on the intestinal walls, or on their contained ganglia, may be proved by the absence of the characteristic symptoms, such as contractions, paralysis, etc., in a portion of the intestinal tube which has been protected from the access of the poison by previous ligation of the branch of the mesenteric artery by which it is supplied. And, conversely, injection of the poison into a branch of the mesenteric artery should, under such circumstances, cause the symptoms first to appear in the portion of intestine supplied by that vessel.]

It should, however, be mentioned that anæmia of the abdominal vessels, as well as dyspnœa, causes an increased peristalsis.

As regards modification of the function of defecation, either diarrhœa, or constipation may be produced by drugs, but as yet it is not known whether the changes are due to action on the motor apparatus of the bowels or on their secretions.[1]

b. ALTERATIONS IN THE SENSIBILITY OF, AND PRODUCTION OF ABNORMAL SENSATIONS IN THE ALIMENTARY CANAL.—Abnormal sensations, nausea, loss of appetite,

---

[1] See Radziejewski, Arch. f. Anal. u. Physiol., 1870, i. [and Hay, Journ. of Anat. and Physiol., 1881 and 1882].

and increased thirst, are very common effects of poisons, especially when given by the mouth, and so brought into direct contact with the sense organs ; the same effects may, however, be often produced when the drugs are otherwise administered. Reliable observation of such effects can only be obtained in experiments on man ; this also applies to the numerous phases of painful sensations which often follow the administration of poisons, such as cardialgia, colic, etc., the causation of which is always obscure.

c. ALTERATIONS IN THE DIGESTIVE SECRETIONS.— Accurate study of such changes, further than the mere evidence of increase of saliva from its flowing from the mouth, or decrease by dryness of the parts, can only be obtained through the production of fistulæ.

[The methods of studying the action of drugs on the salivary and biliary secretion will be given under the heading of the action of drugs on glandular action. Occasionally some idea as to the action on the other digestive secretions is to be obtained by the analysis of gastric and pancreatic juice obtained through fistulæ, and the examination of the digestive products obtained in the same manner. Our ignorance, however, of the conditions, such as nerve-influence, under which these secretions are formed, does not permit of any very accurate studies in this connection ; and very often quite as correct notions may be obtained by adding the drug to artificial digestive fluids.

Dogs are most suitable for gastric fistulæ. The animal is first narcotized with opium or chloroform, bound on his back, and the hair shaved from the epigastric region. An incision is then made through the skin, commencing at the lower border of the costal cartilages, and about an inch and a half to the left of the linea alba, and extending downward parallel to this line, for a distance a little less than the diameter of the flange of the canula which it is desired to use. Each muscular layer is then to be divided in a direction parallel to its muscu-

lar fibres, and every bleeding point tied before opening
the peritoneum. When it is certain that the bleeding
has stopped, the peritoneum is to be opened on a director.
On stretching open the wound, the stomach (which should
have been distended before the operation by a full meal,
or by inflation with air by means of a tube passed down
the œsophagus) comes into view, its oblique muscular
structure being plainly visible through its serous cover-
ing. A point of the gastric wall should now be seized
with artery forceps at a spot where there are not many
large vessels, and drawn forward. Two strong silk
threads are then passed into the walls of the stomach
with a curved needle, at a distance from each other about
equal to the diameter of the tube of the canula, and
brought out at a similar distance from the points where
they were introduced. An incision rather shorter than
the diameter of the tube of the canula, is then made into
the gastric walls between the two threads, and the
opening stretched with blunt hooks until it is large
enough to admit the inner flange of the canula. The
stomach is then tied to the canula by the threads previously passed into its walls, and their ends then passed
through the abdominal walls and tied, thus serving not
only to close the wound in the latter, but also to maintain them in apposition with the stomach. The canula
should be left uncorked for an hour or so after the operation so as to prevent the passing of the gastric contents
into the peritoneal cavity, should the animal vomit. The
dog must be fed on milk for two or three days after the
operation, and kept in a warm place.

The form of canula almost universally used, is that
designed by Bernard. It consists of two silver or
nickel-plated tubes, each of which has at one end a broad
flange; one tube screws into the other, so that the distance between the two flanges may be altered at will.
On the second or third day after the operation, the margin of the wound becomes very much swollen; this
arrangement of the tubes permits the lengthening of the
canula, so that the skin is not ulcerated from pressure

of the flange. The canula may be closed by a cork soaked in a decoction of colocynth, to prevent the dog from tearing it out with his teeth.

Ordinarily the animal will be ready for experiment in about a week. Comparative experiments may then be made on the characters of the digestive process at stated intervals before and after the administration of the drugs, of the changes in the secretion or the drug, and of the absorbability of the poison. Gastric juice can also be collected for experiments on artificial digestion: or infusions or glycerine extracts of the mucous membrane of the stomach in 0.2 per cent. HCl. may be employed.

The action of drugs on the pancreatic secretion is as yet an entirely unbroken field. The extreme susceptibility of the pancreas to disturbing influences will render the study of the action of drugs on its secretion, as obtained in temporary fistulæ, liable to complication, while it is probable that it is impossible to retain a normal condition of the gland in permanent fistulæ.

If it is desired to attempt studies on these points, probably the best method would be to open the abdomen of a dog under warm salt solution, insert a canula in the pancreatic duct, and inject the drug into the gland artery. The method for establishing temporary or permanent pancreatic fistulæ, may be found in Sanderson's or Cyon's Hand-book, or in Bernard's writings.]

Diminution of the secretions may cause dysphagia or constipation, or changes in the digestive processes; increased secretion may produce diarrhœa.

The question as to whether the retained products of secretion in the blood produce further disturbance of function when the secretory processes are interrupted, can only be raised in the case of the bile, and even here it is clouded with a great deal of obscurity. At any rate, the retained substances cannot be regarded as bile, which, as such, is only elaborated in the liver. The possibility of retention of bile through toxic action on the intestinal canal (catarrh leading to obstruction of the bile-duct), should be borne in mind. Such a state of affairs is disclosed by the paleness and abnormal odor of the feces, through the jaundiced color of the skin and certain mucous membranes, and by the presence of the bile acids and coloring matters in the urine. The existence of *icterus gravis* would indicate that the retained bile products may, under certain illy-defined conditions, be the cause of further disturbances of function.

d. ALTERATION IN THE DIGESTIVE PROCESSES.—The presence of poisons in the alimentary canal can lead in the most various ways to digestive disorders; either through alteration in the reaction of the digestive juices, through action on the food stuffs or their digestive products, through preventing the formation of normal secretions, or by action on their ferments, or finally by interfering with the normal fermentative processes. Any one or all of these conditions may be produced, either by directly swallowing the poison or through its passing from the blood into the secretions. The consequences of disordered digestion are first seen in sensory disturbances, as loss of appetite, nausea, or colic; then in motor disturbances, as vomiting, diarrhœa, or constipation, and, when long continued, in emaciation and weakness.

The proof of such toxic changes is best obtained through artificial digestion experiments in which the poison is mixed with the digestive fluids, though occasionally some idea as to the action may be obtained from examination of the vomited matters or feces, or by careful analysis of the symptoms produced.

SECTION IV.—Action on Glandular Organs.

a. SECRETING GLANDS.—The character of the influence of poisons on glandular organs is best made out through study of their secretions; nearly always the changes which will be detected will be of a quantitative nature, and are generally easily enough determined, while qualitative changes produced by toxic action have been but rarely studied.

The mode in which drugs increase or diminish secretions is as obscure as the general physiological processes connected with the normal act. In many cases, doubtless, the act is of a vaso-motor nature, as in the increased salivation produced by curare ; in other cases direct action on the secretory tissues or nerves must be concerned.[1] A thorough investigation is possible in the case of but few glands ; in the case of the salivary glands, however, this branch of pharmacology has been comparatively thoroughly worked out.

1. [ACTION ON THE SALIVARY SECRETIONS.—In order to study the action of drugs on the salivary secretion it is necessary to establish temporary salivary fistulæ in the lower animals, and expose the nerves through whose action modifications of the act of secretion can be produced. Large dogs are the most suitable for such operations.

Since the submaxillary gland is the most accessible it is the gland which is ordinarily selected for such studies.

To make a temporary salivary fistula in a dog, the animal is chloroformed, the hair shaved from the lower surface of the jaw and the side of the neck, and an incision made along the inner border of the lower jaw, commencing about its anterior third and extending back to the transverse process of the atlas, dividing the skin and platysma muscle. After clearing away the connective tissue and fat, carefully avoiding all veins, the submaxillary gland comes into view, just below the angle of the jaw. It is then seen that the gland lies in an angle formed by the junction of two veins which go to make up the external jugular, one branch coming from above downward, directly behind the gland, and usually receiving a small vein from the gland itself (as represented in Fig. 31), while the lower branch runs horizontally below

[1] Heidenhain, Pfluger's Archiv, v. 309.

the gland, and is formed by the junction of two other branches, one coming from above and the other from below ; this horizontal branch very constantly receives a vein from the gland. This dissection requires care, to avoid wounding these large veins.

Fig. 31.

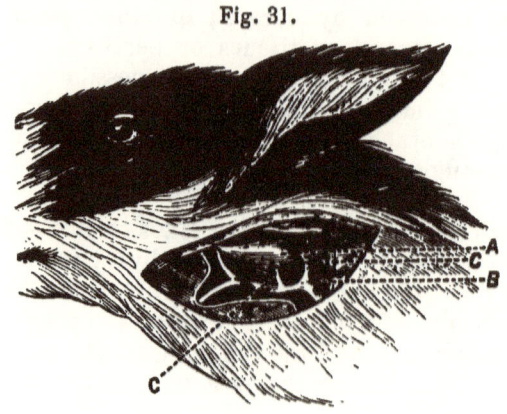

Veins of the submaxillary gland of the dog.   (After Bernard.)   A. Submaxillary gland.   B. Jugular vein.   C. Glandular vein.

Both branches which go to form the horizontal branch are now to be tied, the one coming from above receiving a double ligature, one where it comes over the ramus of the jaw and the other where it joins its fellow, the intermediate portion being removed. Having now carefully removed the cellular tissue from the portion of the wound in front of the gland, the thick belly of the digastric muscle comes into view, its fibres running forward from its origin on the temporal bone, to be inserted in the middle third of the ramus of the lower jaw, immediately in front of the insertion of the masseter, from which muscle it is separated by a slight groove. In front of the digastric the floor of the wound is formed by the transverse fibres of the mylo-hyoid muscle, crossed by the mylo-hyoid nerve, which comes out from under the jaw at the point of insertion of the digastric muscle.

The connective tissue is then gradually to be cleared away, with a blunt hook, from the surface of the digas-

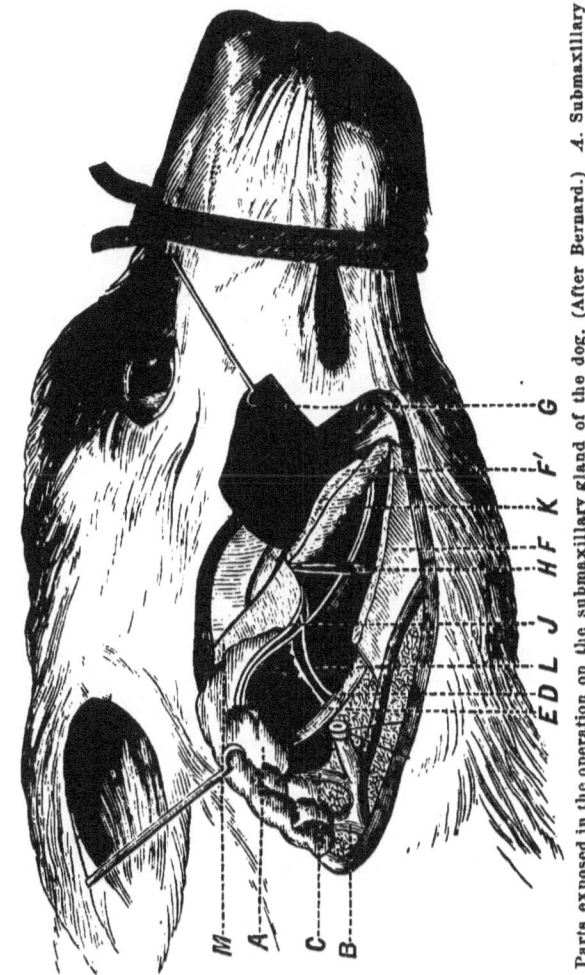

Fig. 32.

Parts exposed in the operation on the submaxillary gland of the dog. (After Bernard.) *A*. Submaxillary gland, turned back so as to show the nerve, artery and duct entering the hilum. *B*. Jugular vein. *C*. Gland- ular vein. *D*. Hypoglossal nerve. *E*. Carotid artery, giving off the internal maxillary. *F. F*. The divided mylo-hyoid muscle. *G*. The anterior half of the divided digastric muscle; the posterior half has been removed. *H*. The lingual nerve, giving off *J*, the chorda tympani. *K*. Submaxillary and sublingual ducts. *L*. Sympa- thetic fibres to submaxillary ganglion. *M*. Masseter muscle.

tric muscle, and from the groove between it and the masseter muscle, taking care to avoid, as the deeper portion is reached, the facial artery, which passes over the jaw to run between these muscles, and the artery to the gland which comes from the facial and goes in this groove back to the gland. In the same locality lie also the ducts of the gland and the chorda tympani nerve. The digastric muscle is now to be separated, with an aneurism needle, from the facial artery, avoiding all the adjacent structures, and its muscular arterial branch tied. The muscle is then divided at its anterior third, or where it is inserted into the jaw, and its posterior extremity seized with a pair of artery forceps, and gradually cleared back to its insertion into the temporal bone, and surrounded by a ligature. Now, when it is assured that there is nothing but muscular structure in the grasp of the ligature, it is pushed back to the temporal bone and tightened, and the digastric muscle divided in front of the ligature and removed. On carefully tearing away the connective tissue at the base of the wound, and drawing back the submaxillary gland, there is exposed a triangular cavity (represented in Fig. 32).

This space is limited above and behind by the deep surface of the submaxillary gland, into the hilum of which enter the artery, chorda tympani, and sympathetic nerve fibres with the glandular duct. Its lower margin is formed by the genio-hyoid muscle, and the upper border by the ramus of the jaw and the masseter muscle ; the anterior portion of its floor is formed by the transverse fibres of the mylo hyoid muscle, on which ramify the branches of the mylo-hyoid nerve.

At the posterior portion of this space the external carotid artery enters and runs along the base of the triangle, giving off first the lingual and then the facial arteries, from the latter of which comes the artery of the gland.

Almost immediately after entering this space the carotid is crossed by the large hypoglossal nerve, running forward to be distributed to the muscles of the tongue, etc. Now, if this nerve is divided at the point where it crosses the carotid, and the central end removed, the pneumogastric trunk comes into view, lying behind the artery. On pulling to one side the vagus trunk, below and behind it can be seen the white trunk of the sympathetic nerve, which here separates itself from the vagus to form the superior cervical ganglion, from which two small filaments pass out to accompany the carotid and glandular artery to enter the hilum. Some of the sympathetic fibres also pass into the gland along the arterial branch which comes from the temporal artery and enters the superior part of the gland.

Then, to expose the chorda tympani and the salivary ducts, the fibres of the mylo-hyoid muscle are to be divided transversely at about their middle, avoiding the nerve and tying all veins, and the upper half of the muscle reflected. The lingual nerve then comes into view, passing from under the ramus of the jaw, and running downward and forward about parallel in direction with the hypoglossal. On drawing the parts toward the middle line, the two salivary ducts are seen passing along close together, immediately under the ramus of the jaw, the submaxillary duct lying nearest the bone and being a little the largest.

On tracing back the lingual nerve to where it passes from under the jaw, it will be seen that a delicate nervous filament here leaves the lingual and curves backward, along with the ducts, to enter the hilum of the gland; this is the chorda tympani. Immediately after the chorda leaves the lingual there is sometimes seen a small ganglionic enlargement, known as the submaxillary ganglion, and as the chorda enters the hilum it forms a slight ganglionic plexus with the fibres of the sympathetic.

Each of the nerves, which it is desired to study, should be carefully isolated and surrounded with a thread, and a canula should be inserted into the submaxillary duct. To facilitate this last step the duct should be freed slightly from the connective tissue and closed with a clip or a ligature, as near the mouth as possible. Then the chorda should be stimulated, so as to distend

13*

the duct with saliva, and a small slip of wood or card passed under it, to act as a support. Now, if one edge of the duct, over the support, is seized by an assistant with a pair of fine forceps, while the operator seizes the opposite edge, and the duct is snipped between the two with a pair of sharp-pointed scissors, the canula can be readily inserted.

The secretion of submaxillary saliva is a reflex act, for which the lingual and glosso-pharyngeal nerves (together with certain other nerves), serve as the afferent fibres, the centre lies in the medulla, while the efferent secretory fibres, together with vaso dilator fibres, pass through the chorda tympani nerve. Drugs may, therefore, cause an increased salivary secretion through stimulation of any of these three divisions of the reflex circle, while the majority of instances in which the secretion is diminished will be found to depend upon paralysis of the chorda tympani.

Thus, when it is found that the injection of a drug into the venous system causes a diminution of the salivary secretion, determined by allowing the saliva to flow from the submaxillary duct into a graduated vessel, it will be ordinarily found (as in the case of atropia) that the stimulation of the chorda fails to produce a flow of saliva ; should, however, it be followed by the ordinary result, increased flow and increased vascularity of the gland, it may then be assumed that the paralysis lies in the centre or afferent nerves. When the paralysis has been located in the chorda tympani, the results of the antagonistic action of some known stimulus of this nerve, such as pilocarpine, should then be tested by injecting a few milligrammes into the carotid artery of the same side, or directly into the duct of the gland.

In most instances it will be found that the paralysis of the chorda can be removed by pilocarpine, and toxic stimulation of the chorda, by atropine.]

2. [ACTION ON THE BILIARY SECRETION.—The action of drugs which modify the amount of bile discharged from the liver may fall under two categories : either action on

the bile-secreting or the bile expelling mechanisms.   It
is probable that these two processes are closely united,
though many instances might be given of drugs, such as
gamboge or magnesium sulphate, which, although powerful intestinal stimuli (and we know that it is from the
stimulation of the intestinal mucous membrane with the
acid chyme that the bile is normally discharged), and
therefore probably possessing the power of causing a reflex contraction of the gall-bladder and expulsion of bile,
cannot be regarded as stimulants to the secretion of bile.
The determination of the point as to whether a drug is
a stimulant of the expelling mechanism is, however, very
much less important than as to whether the substance is
a true hepatic stimulant or not; we will accordingly at
present simply give the methods of examining the action
of drugs on the secretory functions of the liver.   At
the outset, we might say that drugs which stimulate in-
testional secretion usually depress hepatic secretion, and
while drugs which produce slight catharsis only slightly
modify the amount of bile secreted, powerful purgation
produces a marked depression. .The method of studying the action of drugs on the hepatic secretion, as employed by Rutherford,[1] is by means of temporary biliary
fistulæ in curarized drugs.   He has found that when
artificial respiration is maintained in curarized dogs, the
secretion of bile remains tolerably uniform during the
first four or five hours after the commencement of the
experiment, but falls slightly as a longer period elapses.
The composition of the bile remains constant.

    The dog should receive a full meal of lean meat the
day before the experiment, so as to allow of complete
digestion and absorption before the investigation is undertaken.   The animal is then fastened on his back, curarized and artificial respiration maintained, and a glass
canula inserted through an opening in the linea alba into
the common bile-duct, near its entrance into the duodenum, and tied therein.   A rubber tube is then attached

----

· [1] Trans. Roy. Soc. of Edinburgh, vol. xxix. 1879.

to the canula, the gall-bladder pressed so as to expel its contents and fill the tube, and the cystic duct then clamped : the flow of bile can be estimated by allowing it to drop from the rubber tube into a graduated vessel. The wound in the abdomen must be closed, and the animal covered with cotton-wool to prevent loss of temperature. After estimating the rapidity of flow for half an hour or longer, the drug can then be injected into the duodenum, by a syringe with a needle point. It should be mentioned that the bile always flows much more rapidly in the first few minutes of an experiment.]

Wherever the direct contact of the poison with a mucous membrane is found to produce a catarrhal increase of secretion, or when, under similar circumstances, anatomical alterations can be made out in the glands of the mucous membranes, the results may always be attributed to direct actions on the tissues.

3. ACTION ON THE RENAL SECRETION.—Functional disturbances of the kidneys are during life only capable of being studied through the character of the secretion, which may, through the action of poisons, be increased, diminished, or altered in character. These alterations may consist either in the admixture of the poison itself or in the presence, induced by the poisoning, of abnormal ingredients, such as blood-corpuscles, hæmoglobin, albumen, bile matters, sugar, lactic acid, leucin, tyrosin, or finally in mere alterations in the quantitative composition of the urine resulting from modified tissue changes induced by the poison. The kidneys themselves are not always actively concerned in the production of these alterations in the urine. It may, however, be assumed that the kidneys are concerned in diminution or abnormal quantity of urine, which, however, can also be a result of toxic alterations in blood-pressure, when blood-corpuscles or hæmoglobin appear in the urine ; in such cases there exists a toxic inflammation of the kidney parenchyma which is capable of post-mortem demonstration. On the other hand, it appears that toxic alterations of kidney structure, such as are often met with in

fatty degeneration, may exist without rendering their presence at all evident by any alterations in the urine. Such appearances often produce definite effects on the entire organism. The excretion of the poison in the urine can, as already remarked, lead to the entire removal of the poison from the body, and can even render a poison absolutely innocuous. On the other hand, it is conceivable that the occurrence, during the poisoning, of a disordered functional activity of the kidney, may suddenly increase the proportion of poison in the blood, and thus lead to intensified, or new symptoms of poisoning; it is consequently *a priori* probable that a similar train of symptoms would follow the administration of the drug to a system in which the kidneys were already similarly affected. The presence of the poison in the urine may lead to inflammation of the bladder and ureters in the same way that the inflammation of the kidneys may be produced by the passage of the poison through the kidneys into the urine.

Anuria and polyuria, when of extended duration, may produce pathological effects upon the system; uræmia and retention of water in the first case, severe thirst in the other. The presence of abnormal constituents in the urine is only of any general moment when they consist of unoxidized substances, such as sugar and albumen, and therefore entail a loss of nutritious principles.

[The secretion of urine may be increased by all causes which produce an increased blood-pressure in the renal glomeruli: hence drugs may act as diuretics which increase the force or frequency of the heart's beat, which cause contractions of bloodvessels supplying other organs (as the skin), or which cause relaxation of the renal arteries. Thus, profuse renal secretion may be caused by section or paralysis of the renal nerves, from the increased pressure in the glomeruli consequent on the relaxation of the renal arteries; while, conversely, diminished secretion may follow stimulation of the renal or splanchnic nerves. All 'drugs, therefore, which produce increased arterial tension will not act as diuretics unless they at the same time cause relaxation of the renal arteries; thus, when strychnia is injected into the circulation it causes diminution of secretion from constriction of all the arterioles, so acting like stimulation of the medulla; but when the splanchnics or renal nerves are first divided, injections of strychnia then produce increased urinary flow.

In addition, however, to the modifications of renal secretion due to alterations in blood-pressure, drugs may act as diuretics by directly stimulating the renal epithelium.

The rate of urinary secretion may be estimated by opening the abdomen and inserting canulæ into the ureters; the canulæ are then attached to rubber tubes by which the secretion is conducted externally into graduated glass vessels.

To introduce canulæ into the ureters, their lower portion, just before their entrance into the bladder, should be selected. The abdomen may be opened, after emptying the bladder and rectum, by an incision on each side of the recti abdominis muscles or directly opposite the sacro-iliac symphysis, and should be long enough to admit two fingers: when the last of the above-mentioned incisions is employed, the ureters can readily be recognized by the touch at the points where they cross the iliac arteries. Instead of the ordinary straight canulæ, it is better to employ glass or metal canulæ bent at a right angle, the long arm having a length sufficient to extend through the abdominal wound after the short arm has been inserted and bound fast into the ureter; by this means kinking of the tube is prevented.

To study the changes in the renal circulation produced by poisons, either the method employed by Ludwig and Mosso may be used, or the *oncograph* devised by Dr. Roy.

To maintain artificial circulation through the kidney according to Ludwig's method, the carotid artery of a dog is opened and as much blood as possible collected and defibrinated. The abdomen is then opened and a canula

inserted into the renal artery and another into the renal
vein, the vessels being first compressed with clips so as
to prevent the entrance of air.   The kidney is then excised with the greatest possible care and placed in the
warm chamber, the arterial canula being connected with
the flask containing the defibrinated blood, under definite
pressure, and the rapidity of blood-flow from the renal
vein estimated.   The substance being experimented with
is then added to the arterial blood, and comparisons of
this rate of flow made.   This method as employed by
Mosso[1] also permits the estimation of changes in volume
of the organ from varying blood-supply ; but since the
plan pursued by Dr. Roy enables similar studies to be
made on the kidney while still in connection with the
natural blood-supply, it is to be preferred as less liable
to error.   Dr. Roy[2] has found that the degree of expansion of the bloodvessels of the kidney furnishes an extremely reliable idex as to the secretory processes going
on in the organ.   His method consists in inclosing the
kidney, after its exposure through an incision in the lumbar region, in a rigid metal box, of appropriate shape,
containing oil, and of such a construction that while no
hindrance is offered to the entrance or exit of blood by
the renal arteries or veins, any change in the volume of
the organ causes a rise or fall, corresponding in extent,
of a recording lever writing upon the moving paper of
the kymographion.   The number of drops of urine which
fall from a canula tied into the ureter can also be recorded on the same paper, by allowing each drop as it
falls to close an electric current flowing through the
bobbins of an electro-magnetic signal.   In all such experiments, the blood-pressure must be also recorded, otherwise serious errors may be made in drawing conclusions
as to the nature of various changes in the volume of the
organ.   For recording changes in the volume of the kidney, two separate instruments are employed, the one in-

[1] Ludwig's Arbeiten, 1874.
[2] Journ. of Phys., Jan. 1828.

closing the organ and the other for recording graphically the changes in its volume. The form of box which Dr. Roy employed in his investigations on the spleen[1] is equally suitable for studies of changes in volume of the kidney. It consists of an elongated sheet-metal box, composed of two symmetrical halves which are joined together by a couple of hinges. Each of these halves is composed of an outer and inner shell, the latter of which fits accurately into the former, and the two are capable of being firmly screwed to one another by means of screws on the upper and lower rounded edges of the box. Between the two shells is clamped the edge of a thin flexible membrane prepared from the peritoneum of the calf. The membrane is so arranged as to form an air-tight chamber which is bounded on the one side by the flaccid membrane, and on the other by the metal wall of the box. In each of the two chambers thus produced, there are two openings, one pair of which is connected with a T-tube, and thereby with the recording apparatus; the other two openings are fitted with small taps, and are simply intended to allow the air to escape, when the chambers are filled with oil after the kidney has been introduced into the box. At the point of junction of the two halves of the box on the side opposite to the hinges is a narrow slit formed by an indent in the edges of the two halves, which slit is intended to permit of the passage of the renal vessels and ureter. The recording instrument communicates with the interior of the two chambers of the box; its principle is similar to that of Dr. Roy's instrument already described for studying the work done by the heart.

In order to make the experiment, the animal is anæsthetized, a canula inserted into the carotid artery for blood-pressure observation, and one into the jugular vein for the injection of the poison. The kidney is then exposed by an incision in the lumbar region, and gently inserted into the box which should have been previously

---

[1] Journ. of Physiol., Jan. 1882.

warmed to the body temperature, and the two chambers of the box, and the tube connecting the box with the recording instrument filled with warm olive oil.]

4. ACTION ON THE SWEAT GLANDS.—It is here worthy of notice, that many drugs may either produce an abnormal sweat secretion, may reduce the normal amount, or may themselves, or their products, pass into the excretion.

[The secretory activity of the sweat glands, like other secretory processes, is largely dependent upon the amount of blood supplied to the organ ; hence drugs which produce vaso-motor paralysis of the skin, will tend to produce an increased secretion of sweat, and conversely, a diminished blood supply to the sweat glands will reduce their secretions. The comparatively recent experiments, however, of Goltz, Luchsinger, Nauroci, Vulpian, and others, show that there are special nervous mechanisms governing the secretion of sweat as complete as had been previously discovered in the case of the salivary glands. When the peripheral end of a divided sciatic nerve in a dog or cat is stimulated with an interrupted current, a profuse sweat is poured out on the ball of the foot. While this secretion, as in the case of the saliva which follows stimulation of the chorda tympani, is to a certain extent governed by the vaso-motor paralysis thereby produced, it is not entirely dependent upon it, as the same results will follow after clamping the aorta, or even in an amputated limb ; while the analogy with the salivary secretion is made still more complete by the fact that atropia will paralyze the secretory fibres of the sciatic, leaving the vaso-motor fibres intact. Then, again, it has been found that there exist special centres in the spinal cord through whose stimulation a reflex secretion of sweat can be produced. Thus, if the central end of a divided sciatic is stimulated, all the limbs, with the exception of the one in which the nerve has been divided, will perspire ; or, if after division of one sciatic the animal is exposed to a high temperature, the sweat will appear in all portions of the body with the exception of the paralyzed limb. The sweat

14

centres are also excited by carbonic acid in the blood; therefore, drugs which cause dyspnœa will tend to increase the secretion of sweat.

Drugs may increase the secretion of sweat either by peripheral stimulation of the secretory nerves, or by direct action on the nerve centres; as yet, we are unable to speak of action on the centripetal nerves. If a poison, when injected into the circulation, produces sweating in *all* the limbs of an animal in whom one sciatic nerve has been divided, it may be assumed that the action is a peripheral one; if, however, the foot on the side in which the sciatic was cut remains dry, it will be probable that the drug directly stimulates the sweat centres.

The antagonism of drugs may also be treated; thus, atropia will check the secretion started up by pilocarpine.]

5. LACHRYMAL GLANDS.—Increased secretion of tears, apart from any direct irritant action on the conjunctiva or nasal mucous membrane (reflex irritation), is to be expected from those poisons which paralyze the vaso-motor nerves, and can be established by direct inspection.

[We are not yet familiar enough with the nervous mechanism governing this secretion to attempt to explain the *modus operandi* of drugs.]

6. LACTEAL GLANDS.—Very little is known as to the influence of drugs in increasing or diminishing the secretion of milk, while it is well known that many poisons are excreted in the milk, and, therefore, give their toxic properties to this secretion.

b. NON SECRETORY GLANDS.—To this group belong the spleen, and, with the exclusion of its bile-making fuction, also the liver. As regards toxic action on the other glands of this group, that is, the supra-renal capsules, thymus and thyroid, nothing is known; while as regards the lymphatic glands, the most that can be said is that when situated in the neighborhood of tissues in a state of inflammation from the action of drugs, they also will become inflamed.

The manifold, but still obscure, connections of the liver with the processes of nutrition would indicate its probably frequent implication in the causation of the disorders produced by poisons, but still all such disturbances, as for example, fatty degeneration, with the single exception of those exhibited in the character of the biliary secretion, are only to be detected after death.

Studies as to the possible effects of the drug on the glycogen, or possibly the sugar, of the liver, should be made immediately after death. The liver must be thrown instantly after death, after chopping up rapidly into a few pieces, into a large quantity of boiling water, prepared beforehand, and it can then, while in the water, be either chopped into fine pieces or rubbed up into a pulp; it is then to be faintly acidulated and filtered, and the filtrate precipitated with iodide of mercury and potassium solution and hydrochloric acid,[1] and after removal of the precipitate, the glycogen precipitated with alcohol. The sugar can also be determined in the first filtrate by Trommer's test.

[Changes in the circulation of the spleen may be studied by Roy's method. (See Kidney.) The method described for determining the character of the circulation in the case of the excised kidney is also applicable to the liver.]   .

Section V.—Alterations in Tissue Metabolism.

All poisons which disturb the normal digestive pro cesses, produce, when the condition of poisoning is long continued, changes in nutrition analogous to those occurring in prolonged fasting or with insufficient food ; that is, loss of weight, absorption of adipose tissue, paleness of the skin and mucous membranes, loss of strength, and in severe cases, death ; in growing organisms, growth

[1] Brücke, Weiner Acad. Sitzgsber. Math. Naturw. Cl. 2. Abth. lxiii. 1871, Feb. 3.

ceases ; the generative function is suspended, and in the pregnant condition abortion may occur. Similar results in the nutritive functions may also be caused in cases of chronic poisoning without any clear disturbance of digestion being detectable.

In addition to these general disturbances, single nutritive functions may be interfered with as a consequence of the action of poisons ; from the uncertainty surrounding our knowledge of the physiology of nutrition, many of these cannot be explained, and we will here only allude to the best known.

1. ENERGY OF THE ANIMAL OXIDIZING PROCESSES.— The energy of the oxidizing processes occurring in the animal economy is estimated either by measuring the quantity of oxygen consumed or of the two principal products of oxidation, carbonic acid and urea. For the estimation of these gases the methods have already been given, while urea may be estimated by Liebig's method of titration ; on account of unavoidable errors, however, connected with this method, it is better to estimate the total quantity of nitrogen in the urine and feces.[1]

An increase of oxidation processes, as occurs in fever, may, also, very probably, occur in toxic fevers. On the other hand, certain substances, such as arsenic, can diminish the excretion of the products of oxidation ; for none of these can any clear explanation be given.

As general effects of toxic changes in these processes of oxidation we may have, when increased, elevation of temperature, rapid emaciation, and loss of strength ; when decreased, reduction of temperature, deposit of fat and excretion of sugar in the urine. The increase of temperature in the cases of toxic fever may, as in ordinary fever, depend upon other causes than increased oxidation. On the other hand, the deposit of fat and the excretion of sugar is not constantly associated with diminished oxidation, but may depend upon other causes.

[1] Segen, Zeit. f. Analyt. Chemie, 1864, p. 155.

2. DEPOSIT OF FAT IN THE BODY.—Chronic poisoning with many substances causes an increase in the normal amounts of adipose tissue in the body (in the subcutaneous tissue, peritoneum and pericardium), and, in addition, fatty degeneration of various organs, especially the muscles and certain glands, such as the liver and kidneys; the latter can also undergo acute fatty degeneration from the action of certain poisons when continued only for a few days, or even according to some authors, for a few hours. Deposit of fat, with the exception of increased panniculus adiposus, can only be recognized after death. The cause of such changes is entirely unknown, and probably varies in different cases; diminution of the general oxidation processes and inflammatory irritation of the fatty parenchyma have been advanced as possible grounds, but no sufficient proof of their causative action has been given. General muscular weakness will follow fatty degeneration of the muscles, and when occurring in the heart the pulse will be weakened and life endangered not only in this way, but also by the possibility of rupture of the heart.

3. DIABETES.—A number of poisons cause the excretion of sugar in the urine, a condition which can be easily recognized by Trommer's test. Either an increased sugar formation in the liver, or a decreased consumption of sugar in the system will cause diabetes. As the latter state of affairs implies diminished oxidation, the presence of diabetes is a symptom of this derangement of the general nutritive functions, and may be caused by any poison which interferes with respiration, or even, though in both cases inconstantly, by mechanical obstruction to respiration.[1] One of the most remarkable forms of toxic diabetes, that which occurs in curare poisoning, cannot be explained in either of these ways, since curare does not cause an increase either of glycogen or sugar in the liver, nor can any reason be given why artificial respiration, or muscular contractions pro-

[1] See literature of this subject in Arch. f. Path. Anat., xlii. 1.

duced by direct stimulation should in curare poisoning interfere with the destruction of sugar.[1]   From the physiological point of view, the results of Pavy, Tscherinoff, Dock, and others have necessitated a modification of the usual explanation of diabetes, so as to render the view probable in toxic diabetes, that there is either some change produced in the functions of the liver, whereby the change of sugar into glycogen is prevented, or by which the glycogen in the liver and also in the muscles is turned into sugar.   It is, moreover, conceivable that toxic diabetes may be due to some process without any physiological analogue, whereby sugar is formed from some other sources than those recognized as physiological.   For example, such a result might exist as a special action on certain tissues, either directly produced or indirectly through circulatory or respiratory changes. Indeed, it is not improbable that poisons which cause vascular paralysis may produce diabetes by modifying the blood supply of the liver.

When, therefore, it has been established that a certain drug causes diabetes, the animal must be allowed to fast for several days, so as to free the liver from its store of glycogen, and the poison then given; if in such a case diabetes still is found, it may be concluded that it was not entirely due to the modifications of the glycogen of the liver.   Then it must, also, in all cases, be established whether the poison paralyzes the vascular system.

A thorough study with prospect of positive results will in many cases be impossible.

Section VI.—Alterations in the Reproductive Functions.

The general sexual functions, the secretion of spermatozoa in the male, and ovulation and menstruation in the female, are so intimately connected with the general

---

[1] Winogradoff, Arch. f. Path. Anat., xxvii. 533.

nutritive condition of the system, that no profound or long-continued departures from normal nutritive activity can be caused without being reflected on the generative functions.   In addition to this, many drugs act directly on the sexual apparatus, either increasing or depressing its activity.   As yet experimental pharmacology has only dealt with the action of drugs on uterine contractions.

Poisons which cause contraction of the uterus may, during pregnancy, induce abortion, and during labor accelerate delivery.   Abortion, following the administration of a drug, cannot invariably be referred to the direct toxic induction of uterine contractions, since the nutritive changes alluded to above may also induce abortion " from weakness," or abortion may result from the death of the fœtus through absorption of the poison from the maternal system.

In the case of every poison which induces uterine contractions (a condition which can be best studied by opening the abdomen in young non-pregnant rabbits, in which the uterus is normally motionless), it must be determined whether the poison acts directly or reflexly on the motor apparatus, and in the former case, whether the uterine muscles, nerves, or centres are stimulated ; and as the uterus contains several centres, with which one the action is concerned : and, finally, whether the uterine centres are directly stimulated by the poison or reflexly through increased venosity of the blood.   An analogous series of inquiries will arise in the case of drugs paralyzing the uterus, a condition which offers even less hope of thorough elucidation than in the case of uterine stimulants.

[It is generally admitted that toxic uterine contractions are, in the majority of cases, due to contractions of the arterioles, either supplying the uterus or the brain ; hence the uterine and cerebral and spinal nervous mechanisms are only indirectly stimulated by the drugs, the effects being primarily due to changes in blood-supply. When contractions fail after section of the cord after the administration of a poison which otherwise is capable of inducing uterine contraction, it may be assumed that they were produced by circulatory changes in the brain.]

The action of drugs on the development of the egg has not been studied, though it is known that poisons, in the case of birds, may pass into the ovum while still within the ovary; or into the developing egg or embryo, by means of the placenta, while within the cavity of the uterus.

Section VII.—Alterations in Temperature.

The alterations of body temperature, frequently produced by poisons, may depend upon either a modification of heat-production or of heat-loss; in both cases the normal regulating processes must be interfered with. Increased loss of heat may occur when the cutaneous bloodvessels are paralyzed, or diminished loss of heat when contracted; in nearly all other cases it will be found that the heat-producing functions are at fault. The muscles are the organs mainly concerned in heat-production; so when a poison causes convulsions the temperature will usually be raised by increased heat-production, and muscular paralysis will lead to lessened heat-production, with the appropriate changes in the body temperature; unless the losses of heat also suffer some disturbance which can nullify these results, and maintain a normal standard.

[For methods of making calorimetric studies, see Sanderson's Hand Book to the Physiological Laboratory, or Prof. H. C. Wood's Memoir on Fever.]

Section VIII.—Action on the Muscles.

Paralysis and abnormal muscular contractions are often produced by the action of poisons; whether these depend upon direct action on the muscle, or through the medium of the nervous system, can readily be determined

by simple experiments. In paralysis it is only necessary to stimulate the muscles directly, and if contraction
occurs, the trouble is proved to be in the nervous system ; in convulsions a positive result may be obtained by
section of the nerve of any selected group of muscles ;
for if the convulsions persist, they must be due to some
cause acting either directly on the muscular fibres or on
the intra-muscular nerve endings. In order to discrimi-
nate between these two possibilities, the animal must be
poisoned with curare, which paralyzes the intra-muscu
lar nerve fibres (of course in such cases in warm-blooded
animals artificial respiration must be maintained), before
the administration of the drug under study ; if, under
such circumstances, the convulsions do not appear, it
may be confidently concluded that the intra-muscular
nerve fibres were the seat of the stimulation, while their
appearance would indicate action on the muscular fibre.
Indeed, as will be shown presently, it can often be determined from the general character of the convulsions
whether they are of central or peripheral origin, and in
cold-blooded animals the methods of exclusion mentioned
on page 32 may be employed. In warm-blooded animals, a poison which produces its results through direct
action on the muscles or peripheral nervous system,
whether stimulant or depressing, will first produce its
characteristic action after hypodermic injection in the
parts with which the poison comes in contact, so showing
its peripheral mode of action. More extended experiments as to the changes in irritability of muscles, as well
as alterations in energy and time of contraction, are best
made on the excised frog's muscle, in the manner described on page 31.

In many cases poisons will cause alterations in the
normal positions of the limbs, from paralysis or spasm
of individual muscles.

Under this heading may also be considered the action
of drugs on the pupil.

1. ACTION ON THE PUPIL.—[The size of the pupil depends upon the degree of contraction of the two antago-

nistic muscular systems of the iris, the circular constrictor fibres, supplied by the oculo-motor nerve, and the
radiating dilator fibres, supplied by nerves derived from
the sympathetic system. Both sets of nerves are normally in a state of constant excitation, since if the oculo-
motor is divided or paralyzed the pupil dilates, and if the
sympathetic is divided in the neck, or its terminal fibres
paralyzed, the pupil contracts. In addition to changes
in the size of the pupil dependent upon direct irritation or
paralysis of these nerves, the pupil may be contracted,
(1) reflexly through the oculo-motor nerve by stimuli
applied to the optic nerve, which thus acts as the afferent nerve; (2) when the eye is accommodated for near
vision ; (3) when the eyeballs are rotated inwards. The
pupils are dilated, (1) during dyspnœa through irritation
of the cilio-spinal centre, the irritation being transmitted
through the sympathetic nerve to the dilator fibres of the
iris ; this dilatation ceases when the dyspnœa passes into
asphyxia, and does not occur after division of the cervical sympathetics ; its production in this manner by drugs
can be therefore readily excluded ; (2) during powerful
irritation of sensory nerves; and (3) during violent
muscular movements.

In addition, still, to these central mechanisms, there
appears to be present in the eye itself some apparatus
by which dilatation or contraction may be produced :
for when the third nerve is divided, and the pupil dilated
under the full influence of the sympathetic nerve, the
instillation of atropia still further dilates the pupil ; and
when physostigmine, under the same circumstances, is introduced into the eye or system, the pupil is contracted.
It is probable, therefore, that the dilator and constrictor
nerves of the iris produce changes in the size of the
pupil by the transmission of stimuli to some local apparatus seated in the eye itself.

To determine the character of the action of drugs on the
pupil, the effects of both local application and venous injection should be studied : the positive decision, however,
as to the precise means by which the toxic mydriasis or

myosis, as the case may be, is produced, is a matter of the greatest difficulty, the *modus operandi* of even such well-known drugs as atropia and eserine being still a matter of controversy. At the outset, reflex causes of change in the pupil, such as are produced by dyspnœa. etc., must be excluded. The question will then arise as to whether dilatation is due to paralysis of the oculo-motor or excitation of the sympathetic, or both, or whether the action is purely local on the eye. Constriction of the pupil will of course be due to the opposite conditions.

Let us first assume that we are dealing with a drug which produces dilatation of the pupil; the first point then to determine is whether the sphincter muscle of the iris still preserves its capability of contraction. This may be determined by the method employed by Bernstein and Dogiel;[1] four wires are connected with the poles of the secondary coil of an induction apparatus and their free ends (arranged in the form of a square, similar poles occupying diagonal corners) placed on the inner edge of the iris. By this means the circular fibres of the iris will be stimulated, and if they retain their normal functions, the pupil will contract. If no change occurs in the size of the pupil when this experiment is properly performed, it may be assumed that the drug produces dilatation of the pupil by paralysis of the sphincter muscle, and no other experiments need be made; in the great majority of cases, however, the pupil will still contract, and it is then necessary to determine whether the dilatation is due to paralysis of the oculo-motor nerve, or spasm of the dilator muscle.

In order to test the irritability of the oculo-motor nerve, it is necessary to open the cranial cavity and remove the cerebral hemispheres.

To make this experiment, the drug is first instilled into the conjunctival sac in a dog or rabbit, and the change

[1] Verhandl. d. nat. Med. Vereins zu Heidelberg, iv. 28, Hermann.

in size of the pupil noted in millimeters. Tracheotomy is then performed, artificial respiration maintained, and both carotids ligated in the neck ; the next step is to remove the vault of the cranium with bone forceps and to elevate the cerebral lobes with a spatula ; after division of the olfactory and optic nerves, the oculo-motor nerves may be found on the sella turcica. The cavernous sinus must be carefully avoided: bleeding from the posterior cerebral arteries may be controlled by slight pressure with a moist sponge. The oculo-motor nerve on the side of the eye in which the poison was instilled, is then to be divided as near the brain as possible and the peripheral end placed on the electrodes of an induction coil and stimulated ; if no contraction of the pupil is produced thereby, it may be concluded that the terminal fibres of the nerve supplying the constrictor muscle of the iris are paralyzed. (For convenience of measurement of the size of the pupil, it is advisable to slit up the external commissure and draw down the lower eyelid by a weighted thread passed through it and the nictitating membrane.)

Should the pupil contract on stimulation, it remains then to be seen whether the dilatation is due to stimulant action on the dilator apparatus ; and even when stimulation fails to narrow the pupil, it is well to see whether there is an associated spasm of the radiating fibres of the iris. This may be determined, as suggested by Bernstein and Dogiel, by placing the two electrodes of the induction apparatus on the side of the cornea; if the pupil dilates still further, the integrity of the radiating muscles is proved, while the functional condition of their nerve supply is tested by stimulation of the cervical sympathetic.

The above experiments would serve to give a somewhat decisive answer to the question as to the *modus operandi* of mydriatics were it not known that many drugs, such as atropia, are capable of producing dilatation of the pupil in the excised eye of the frog, where of course, there can be no question as to paralysis of the oculo-motor nerve endings which have been already sepa-rated from their nerve centre ; and even after division of the oculo-motor nerve (for methods see Cyon *Methodik*, p. 510, or Bernard, *Physiologie Exp. de la Sys. Nerv.*) atropia is capable of still further dilating the pupil.   In view of such facts it is necessary to assume the existence of peripheral ganglia in the iris itself, where indeed nerve-cells have been found in abundance.

In the case of drugs which produce contraction of the pupil (myosis), the probability will lie in favor of a spasm of the sphincter muscle when there is also found by the ophthalmoscope to be spasm of the muscles of accommodation.   Additional evidence will be added to this view when the drug produces a higher degree of contraction than follows section of the sympathetic, when atropia is capable of overcoming the myosis produced by the drug, and when irritation of the sympathetic, in an eye so treated, is capable of still further dilating the pupil.]

Section IX.—Action on the Nervous System.

In previous chapters reference has already been made to several derangements of the nervous system produced by the action of poisons, but it still seems advisable to again group them under their appropriate heading.

A. Action on the Organs of Conduction.—Toxic action on the nerve trunks occurs much more seldom than on the nerve terminations.   This may probably be due to the limited supply of blood which these portions of the nervous system normally receive, since they are equally sensitive to direct action with solutions of the poison (see p. 50).

In the case of many poisons which act primarily on the nerve-endings, particularly in vascular organs such
15

as the muscles, the functional disturbance gradually spreads up the trunk of the nerve.

The functional disturbances produced in the organs of conduction by the action of poisons, may be of two kinds: first, modifications in the normal nervous stimuli; second, in abnormal stimuli. The first of these conditions may be manifested in diminution or abolition of irritability (loss of the power of conducting impressions), or in exaltation of irritability or in abnormal persistence of excitation. Loss of the power of conduction in nerve trunks is disclosed by the absence of the usual result of stimulation, as, for example, the non-appearance of vascular contraction after normal, central, or artificial stimulation of a motor nerve, and in the case of sensory nerves, in abolition of sensibility in the surfaces on which these nerves are distributed.

[It should be noted that while but few drugs completely destroy the power in nerves of conducting impressions, many poisons will depress the irritability of nerves below normal. When, therefore, it is found that after death both the muscles and nerves are capable of responding to stimulation, it should then be tested whether there has been any alteration in the degree of irritability. To accomplish this, the vessel in one hind leg of a frog is ligated, and the poison injected under the skin of the back. When the effects of the poison are clearly marked, the sciatics are laid bare, and the strength of current which is necessary to produce contraction determined for each limb: the irritability of the poisoned and non-poisoned sciatics can be thus compared.]

Increase (accumulation) of the irritant action during its conduction along the nerve, has never been observed as a result of toxic action; but, on the other hand, abnormal persistence of stimulation, such as the production of tetanus or prolonged muscular contraction, as a result of a single stimulus, has been described in numerous instances. Abnormal irritations evoke convulsive muscular contractions where the motor nerves are concerned, and painful or abnormal sensations, or subjective sensations for the nerves of special sense, when the stimulus affects a sensory nerve.

The first point to be determined in the study of toxic effects of the above nature (convulsions, paralysis, etc.), is whether they are due to alterations in the organs of conduction. In the case of paralysis, this is readily determined by separate stimulation of nerve and muscle. If the latter (direct stimulation) produces a muscular contraction, while indirect stimulation does not, it may be concluded that there is some interruption in the conductivity of the nerve, and if the stimulus remains ineffective, even when applied to the nerve immediately before its entrance into the muscles, it would indicate that the loss of function lies in the intra-muscular portions. To decide in such a case whether the remainder of the nerve is implicated, the muscle ·may be excluded from the poisoned circulation in the manner described on page 32, or galvanic stimuli employed. If convulsions are produced, their nature will give some general idea as to their origin. Convulsions of central origin throw the entire muscle, and usually entire groups of muscles, into co-ordinated contractions ; on the other hand, convulsions originating in the nerves, or in their intra-muscular endings, usually implicate different muscular fibres at different times, and with different degrees of vigor, producing the so-called " fibrillar contractions."

[Even fibrillar contractions, however, may be of central origin, since Von Anrep has found that after injections of nicotine they occur in limbs protected from local action of the poison by ligation of their bloodvessels.]

Above all, however, it should be remarked that such contractions persist after separation of the nerves from the central organs, a procedure which will interrupt or prevent the appearance of convulsions of central origin. Finally, convulsions due to action on the organs of conduction do not appear in limbs which have been protected from access of the poison by ligation of their blood-

vessels, while convulsions of central origin affect equally
all members, whether connected with the circulation or
not, provided only that the nerves remain intact. . [Con-
vulsions of cerebral origin disappear after removal of
the medulla, while convulsions of spinal origin persist.]

B. Action on the Peripheral Nerve Endings.—
Sufficient has already been said on page 50, as to the
action of poisons on the peripheral terminal fibres of
motor nerves; at present we are only concerned with
the peripheral endings of centripetal nerves.   These
organs are liable to toxic action whenever a poison,
either by direct contact or by means of the circulation,
is brought to the organs in which they are situated.
The toxic influence may be manifested either by excitation or paralysis, or occasionally the latter condition
may, as a result of the.same drug, follow a condition of
increased irritability.   The consequences of stimulation
of the sensory nerves may be evidenced either by various
abnormal sensations (subjective sensations for the nerves
of special sense), or in modification of reflexes, such as
secretion of tears, saliva, or in vomiting.   Paralysis of
these organs is followed by insensibility and loss of
reflexes.

Such disturbances of function can be readily localized
in either the central or peripheral nervous system.
Symptoms of irritation which disappear on section of the
nerve as near the periphery as possible, and symptoms
of paralysis which are not evident when the stimulus is
applied directly to the nerve-trunk, must be due to the
action of the poison on the peripheral nervous apparatus.
But, unfortunately, the only animals on which such decisive experiments can be made are the very animals
which are most unsuitable for the study of the points in
question; therefore we are practically restricted, as far
as accurate results are concerned, to such evidence of
the state of the conductive sensory function as is found
in the character of the reflexes, and in these experiments
it is almost impossible to distinguish between results due

to action on the peripheral, conducting, or central sensory organs.

The same class of experiments may be made as has been given for the study of the action of poisons on the motor nerves. When the action is localized to the point of application of the poison, or can be excluded from a limb by ligation of its bloodvessels, or when it appears more markedly and rapidly in an organ in whose arteries it has been injected, it indicates that the principal action is on the peripheral organs. If one possessed a means of paralyzing the peripheral endings of centripetal nerves, as curare paralyzes the centrifugal, all phenomena depending on irritation of these structures must be absent after the administration of such a poison, and the symptoms which appear before its administration, and fail to appear after it, must be due to stimulant action of the peripheral sensory apparatus; unfortunately, no such drug has yet been discovered.

[Various methods have been proposed for investigating the condition of the centripetal nerves; all are, however, open to objection. Probably the best method is to ligate the bloodvessels in one limb of a frog, and then inject the poison into the dorsal lymph-sac; after the effects of the poison have been developed, comparisons can be made between the irritability of the central ends of the sciatic nerves; that is, whether a stronger irritant is required on the poisoned than on the unpoisonous side to produce reflex movements. The success of this experiment will depend upon the integrity of the motor nerves, for, if they are paralyzed, of course no reflexes can be produced. In such cases one hind limb of a frog may be partially amputated, leaving only the sciatic nerve intact, and the poison then injected into the amputated limb, and the capability of transmitting impressions tested by comparing the reflexes produced by stimulation of the skin of each hind foot.

Von Bezold proposed a method of testing the irritability of sensory nerves, in which the reflex contractions of a frog, feebly under the influence of strychnia, served as an index of sensory excitation. This method, as modified by Pflüger, consists in the following : the blood-vessels are tied in the lower extremities of a frog, and these limbs then so severed from the body that their only connections are the sciatic nerves. These nerves are then kept constantly moistened, one with a solution of phosphate of sodium, and the other with a solution of the drug, both solutions being of the same degree of concentration. Then by irritating the central end of each sciatic on the distal side of the solution, a general idea can be obtained as to the action of the drug on the sensory nerves. The objection to this method is the difficulty in comparing the osmotic equivalents, under such circumstances, of different drugs. In the case of drugs which have been proven to be inactive on the motor nerves, the first method is the best, even when the drug is known to modify the functions of the spinal cord.]

Another question, which also is difficult to answer positively, is whether the results above alluded to are due to specific action on the nervous apparatus, or to some action on the tissues in which they are located, as it is well known that every strong mechanical or chemical lesion of a tissue acts as an irritant to the nerve-fibres of that organ. It is only possible, however, to express an opinion as to this point when the organ is the seat of gross anatomical changes ; when these are not to be detected, we must conclude that the drug exerts a specific action on the nerve endings. But even should anatomical lesions be found, the possibility of a specific action on the nervous organ is by no means excluded ; for these changes, depending upon disturbances of circulation, leading to hyperæmia and inflammation, may themselves be the palpable reflex result of action on the nervous system. A specific action on the nervous system may be assumed when sensory or reflex phenomena (pain, vomiting, sneezing) appear without any evident change, or with only tardy alteration in the tissue on which the drug is acting. A specific action can be further inferred when a drug produces different results when applied to different tissues of the same general functions, but with different nerve-supply.

C. ACTION ON THE CENTRAL NERVOUS SYSTEM.— Action under the most manifold forms on the nerve centres is one of the most frequently observed effects of poisons. As the action on the nerve centres of the heart and intestine has been already studied, the present chapter will be confined to the consideration of toxic action on the brain and spinal cord.

The separation of the action of poisons on the central system from that on other organs is usually easily attained. In a large group of these phenomena, the sensory, no doubt, can arise; and further, irregularity in a rhythmical motion can only depend upon action on a nerve centre, though alterations in the frequency of rhythm can be caused by action on the centripetal regulating nerves, a possibility which is readily excluded by section of those nerves. But in many other cases, the toxic phenomena are so manifestly of central nervous origin that no such control experiment is necessary. Co-ordinated contractions, such as the movements of deglutition; paralyses of muscular groups whose movements are governed by a special nerve centre, such as the muscles of respiration; painful sensations or anæsthesia of the entire skin, or of surfaces in no direct communication with the point of absorption of the poison, are all, with great probability, to be referred to specific action on the nerve centres. Still, some limitation of this statement should be made for the group of motor phenomena, since many of the co-ordinated muscular movements, such as deglutition and vomiting, can also be produced by toxic stimulation of certain peripheral organs.

The cause of phenomena resulting from exaltation of function, such as convulsions or sensory phenomena resulting from toxic action, can be localized in the central apparatus by section of the nerves going to the affected organs, when of course the disappearance of the symptoms would prove their central origin. In certain co-ordinated movements, the possible peripheral origin may be excluded by section of the appropriate nerves. Motor and sensory paralyses, whose general character fails to give any indication as to their origin, must be regarded as central when the peripheral origin has been excluded by the means mentioned in the preceding chapter.

In all toxic phenomena whose central origin has been established by these methods, the question arises as to whether they are due to direct specific action on the centre, or to indirect action through changes in the blood, or to local disturbances of circulation in the nerve centres by action of the drug on the heart or vessels; since any one of these departures from the normal relations will exert a profound influence on the nerve centres. Such points may be readily investigated in the frog; for if the same class of symptoms appear in the frog as in the warm-blooded animals they can be safely attributed to specific action on the nerve centres, since we have found that the functions of the nervous system are in cold-blooded animals independent of the state of the circulation. Should, however, the symptoms of poisoning vary in the two classes of animals, it cannot be concluded with the same degree of confidence, though the conclusion will be correct in many cases, that in the higher animals they are due to indirect action on the nervous system, since the results may be attributable to different modes of action on the different species. The supposition as to indirect action will be greatly confirmed when it is known that the drug in question produces alteration in the function of the heart, respiratory apparatus or blood, it only being necessary to determine which group of phenomena first appears. In certain cases control experiments may be made in preventing, or compensating for the respiratory or circulatory action of the poison; for example, changes in the respiratory gases of the blood may be prevented by artificial respiration; contraction of the bloodvessels, by large doses of curare. If, after experiments of this nature, the same symptoms do not result, it may be positively concluded that they are due to an indirect nervous action only.

1. INTERFERENCE WITH THE AUTOMATIC FUNCTIONS.—The automatic functions of the medulla oblongata, especially the innervation of the respiratory movements, the regulation of the heart, vaso-motor tonus and pupil mechanism, are extremely often disturbed by the action of poisons. Many poisons act on all these functions simultaneously, therefore associating them in somewhat the same functional analogy in so far as they are all brought into a condition of excitation by the venosity of the blood in dyspnœa. Poisons may either exaggerate or annul these functions, or modify the frequency of the rhythmical movements under their control. Very often two stages occur in poisoning by one drug, first an increased vigor of function and acceleration of rhythm, then enfeeblement and retardation. The characters and methods of studying these changes have been already given.

With these alterations of functions are intimately associated the motor phenomena of irritation produced directly by the action of the poison, since in the present status of physiology, the so-called automatic central stimulations are regarded as conditions in which an irritant acts directly on the nerve centre, and not the conduction of an irritant through the centripetal nervous system. As examples of this form of toxic stimulation may be mentioned vomiting, when not produced by toxic action on the peripheral nerve endings, intestinal and uterine contractions from action on the brain and cord, and general convulsions from action on the so-called "convulsive" centre in the medulla; certain of these conditions may be produced by a high degree of venosity of the blood, a fact which should be remembered in forming an opinion as to the cause of the phenomena.

2. REFLEX AND CO-ORDINATED FUNCTIONS.—In the normal state, the centripetal impressions produce orderly reflexes, which are capable of being controlled or prevented by automatic inhibitory centres in the brain or

by the will. Poisons can modify these phenomena in the following different ways: *a*. The limitation of the reflexes to certain single normally associated motor apparatus can be so suspended that every centripetal impression throws the entire mass of centrifugal fibres into excitement, thus producing general convulsions; these reflex convulsions are ordinarily tetanic in character, and each part of the body assumes the condition which must follow from the simultaneous contraction of all the muscles associated with it. Hence, in such states the back is hollowed (opisthotonos), the head extended, the jaws tightly closed (trismus), and the limbs extended. In weaker degrees of such action, as well as in initial stages of violent action, the reflexes are only abnormally increased in strength and in the number of associated muscles, so that the character of co-ordinated movements is lost. These convulsions differ from those described in the preceding chapter, in that they can only be inaugurated by centripetal impressions, for which, however, the lightest touch or jar will often suffice.

The removal of all forms of external stimuli can only be accomplished, with any degree of certainty, in the case of frogs, by placing them under a bell-jar on some immovable support.

b. On the other hand, poisons can weaken or entirely prevent the production of reflexes, when the animals lie absolutely insensible to all forms of stimulation: ordinarily this condition follows the state of affairs described under *a*.

c. The reflex inhibitory apparatus can be brought into either a condition of increased or diminished functional activity. This will be considered directly.

[In order to study the action of drugs on the reflex functions, the method of Türck is probably the best: its principle consists in comparative measurements of the time required before and after poisoning for a given stimulus applied to the skin to evoke a reflex muscular contraction in a frog from whom the cerebral hemispheres have been removed.

In order to determine the state of the reflex functions of the spinal cord, the brain and medulla are separated from the cord by an incision made through the occipito-atlantal membrane: this locality may be readily recognized by the touch as a depression lying in the median line of the back on a line drawn across the skull at a tangent to the posterior borders of the membrana tympani. To divide the cord, the frog is held in the left hand, and the head strongly flexed on the neck by the left thumb ; an incision, a few millimetres in length, is then made with a sharp-pointed knife through the skin and membrane, care being taken not to extend it too far to the sides, when, on removing the blood with a sponge, the medulla will come into view, if the head is kept well flexed, and may be divided. This method is preferable to one thrust of the knife, or to the division of the cord with the scissors, when it can never be known whether the section is complete or not.   The brain is then to be destroyed by breaking up the contents of the skull with a needle. The hemorrhage will usually soon cease : should it continue, it may be checked by the insertion of a small plug of wood into the opening in the skull.   After the operation, the frog should be placed under a moist bell-jar for about an hour, until the shock of the operation has passed off.   At first the limbs will all be extended, and probably no motion can be produced by irritation, but after a while the limbs will be drawn up and a more nearly normal attitude be regained, marking the returning tone of the spinal ganglia.

When it is believed that the shock has entirely passed off, the frog can be suspended by a tack or hook passed through its nose to some suitable support, care being taken that no part of the frog's body comes in contact with any solid.   Draughts of air must be avoided, and the skin must be prevented from drying by frequent immersions in a basin of water.   A test liquid is then made by diluting with water 1 c. c. of sulphuric acid to a litre ; a few cubic centimetres of this acid are then placed in a small glass beaker, and the glass brought under the frog and elevated until the tip of the longest toe just dips below the surface of the acid, and the length of time which is required before the frog withdraws his foot determined by counting the beats of a metronome. The instant the reflex movement occurs the entire foot must be washed in a large beaker of water to remove the excess of acid and prevent corrosion of the skin of the foot. The time elapsing between the immersion and withdrawal of the foot is then to be written down, and after waiting five minutes, the experiment repeated, care being taken to immerse the same toe to precisely the same extent in each trial. These experiments are to be repeated until three successive trials give about the same numbers, when the drug can be injected and the time of reflex action compared with the normal standard. It is probable that most substances, which are at all irritating, will at first, from stimulation of the sensory nerves, reduce the spinal reflex irritability, or, in other words, lengthen the time of immersion.

Every reflex action requires the functional activity of a sensory nerve, nerve centre, and motor nerve; consequently, before the activity of the spinal ganglia can be studied, the sensory and motor nerves must be known to preserve their functions; therefore, studies of the action of the drug on the peripheral nervous system and muscles, should precede the examination of the condition of the spinal cord. Should, however, for any reason, this not have been done, and it be found that after the administration of the drug, reflex action gradually disappears, the condition of the motor nerves must be tested with a weak induction current; if they preserve their irritability, it may be concluded that either the sensory nerves or spinal centres are paralyzed. The latter possibility may be excluded by first ligating the bloodvessels of the limb which it is proposed to stimulate, thus protecting the terminations of the sensory nerves from the poison; if the reflex irritability is now depressed, it may safely be attributed to action on the nerve centres.

The same method may be employed in studying reflex

action in the case of drugs which are known to paralyze the sensory nerves ; while in the case of drugs which paralyze motor nerves or muscles, the limb opposite to the one stimulated may be protected by ligation.

Having determined the action of the drug on the spinal cord, the question will now arise as to its action on the cerebral inhibitory centres of reflex movement. It has been discovered by Setschenow that the optic lobes in the frog contain centres, stimulation of which depresses the reflex activity of the spinal cord, while their removal exalts it ; though no doubt can exist as to the accuracy of these facts, considerable controversy still exists as to their explanation, but until more conclusive proof is brought forward as to the falsity of Setschenow's theory, we will accept with him the doctrine of special spinal inhibitory centres in the optic lobes, whose activity is capable of being stimulated or depressed by various agents. In order to study the action of drugs on these centres, all portions of the cerebrum, anterior to the optic lobes, must be removed by section with scissors through the skull, on a line with the *anterior* margins of the tympanic membranes. After observing the precautions mentioned above, the normal degree of reflex irritability is determined and the drug administered ; should the activity be depressed, the medulla is then divided on a line with the posterior margins of the tympanic membranes, and the reflex activity again tested. If it is then found that the reflex functions are greatly exalted, it may be concluded that the initial depression was due to stimulation of Setschenow's centres. Or, on the other hand, drugs may paralyze these inhibitory centres and so bring the reflex activity of the cord, even when in connection with the medulla, up to the normal degree of the isolated cord. In the removal of the cerebrum, care must be taken to avoid hemorrhage, which is a stimulant to Setschenow's centre, as is also a decreased action of the heart. It has also been urged by W. T. Sedgwick,[1] that the depression

---

[1] Journ. of Physiol., vol. iii. No. 1.

of reflex action following the administration of quinine is
due to stimulation of the afferent fibres of the vagus
nerve, an inhibitory effect on the spinal centres being
produced in the same way as when any afferent nerve is
stimulated.]

In addition to these general disturbances of reflex
action, poisons can also influence individual reflexes; but
since these reflex centres are, in all probability, identical
with the co-ordinating centres, it cannot ordinarily, as in
spasm of the muscles of deglutition, be determined whether
the toxic action is exerted directly on the centres of co-
ordination, or whether their reflex stimuli are only abnormally increased. The position is here much the same as
in the case of general reflex convulsions (strychnia
tetanus), where the question whether the spasms are
evoked by external stimuli can only be settled on the
frog, and not on the warm-blooded animals.

3. ACTION ON THE SENSORY FUNCTIONS.—As already
stated, toxic disturbances of the sensory functions can be
studied with any degree of certainty only on man.   The
class of drugs which can be thus examined is consequently
extremely limited, adding still another difficulty to an
already obscure subject.

These toxic sensory phenomena consist of two consecu-
tive stages: 1. *The stage of excitation, with its consequent phenomena.*   *a.* The sensory perceptions no longer
preserve their normal relation to the objective stimulant,
but exist in relatively increased intensity, thus producing
an increased sensibility; errors in the estimation of the
character of the stimuli may also be produced, and in
marked toxic conditions sensory peceptions (subjective)
may be produced without apparent cause.

b. In slight degrees of this form of toxic action the
flow of ideas is facilitated and the mental processes stimulated, though mental control is diminished; in severe
forms of intoxication the mental processes are entirely
uncontrollable and disorderly, giving rise to the various
forms of toxic mania and hallucinations.

c. In slight degrees of poisoning the normal control

of the voluntary movements is lost, the muscles, perhaps from disturbance of the muscular sense, contracting more or less powerfully than was intended, thereby producing various disturbances in locomotion or speech, while in severe forms of intoxication the power of performing normal movements is entirely lost, and complete paralysis or general or localized convulsions result. The cerebral stimuli of movement may also be disturbed and lead to various maniacal acts; the highest forms of these sensory disturbances constitute delirium.

2. *The stage of depression* is characterized by phenomena opposed in every respect to those just mentioned; in moderate degrees of intoxication, sluggish senses, obtuse mental acts, and indisposition to move; in higher degrees, complete loss of consciousness (sleep), and depression of all the reflexes, with impossibility of being aroused (sopor, coma, narcosis). In the case of many poisons these last-named phenomena may precede the above mentioned, or exist alone as special actions of the drug.

Another sensory disturbance, less often produced by drugs than by other causes, is syncope; it consists of loss of consciousness without the symptoms described above as constituting the initial stage of excitement.

The only premonitory symptoms of syncope are found in disturbances of special senses (darkness before the eyes) and disturbance of muscular co-ordination (dizziness, staggering), appearing before complete loss of consciousness is established.

The toxic sensory disturbances may depend either upon direct action on the cerebrum or upon disturbances of respiration or circulation, both of which are, however, functions essentially under the control of the central nervous system. Disturbance of respiration may merely produce loss of consciousness, and that only in the stage of asphyxia through insufficient supply of oxygen; while circulatory disorders may produce the most manifold sensory symptoms as soon as the blood-pressure in the cerebral vessels suffers any considerable change; thus syncope may be caused by depression of the blood-pressure from diminished vigor of cardiac contraction. The effects of increased arterial pressure and of venous stagnation are less clear ; both conditions, of which the first alone can be regarded as a direct toxic result (from paralysis of the cerebral bloodvessels, increased force of the heart's action, or contraction of the peripheral vessels), are usually designated as congestion of the brain, and since post-mortem examinations in cases of poisoning accompanied by sensory phenomena often reveal positive changes, such as hyperæmia of the brain and its membranes, the anatomical conditions are supposed to be a cause of the symptoms. This hyperæmic condition of the brain is often manifested during the action of the poison by congestion of the face, injection of the eyes and increased secretion of tears and saliva ; while in many cases even actual rupture of the vessels (cerebral apoplexy) may be produced.

Though it is acknowledged, as is proved by the symptoms occurring in cerebritis or meningitis, that hyperæmia of the brain may cause symptoms similar to those described above, there is no means of proving absolutely, that, in the production of these symptoms by poisons, the same processes are concerned; accordingly, unless the contrary can be proved, it is not unwarrantable to ascribe the sensory toxic effects of drugs to direct action on the cerebral centres.

The proof of circulatory disturbance in the brain is only obtained with the greatest difficulty. The post-mortem appearances are apt to be deceptive, since it is conceivable that transient circulatory changes may exist without leaving any characteristic post-mortem appearances, and congestion etc. (except, of course, active inflammation), may be produced during the death struggle. During life, observation of the retinal vessels with the ophthalmoscope will give a tolerably accurate idea as to the state of the cerebral circulation. Direct inspection in animals is rendered possible by trephining, a glass plate being inserted, after the method of Donders, in the opening in the skull;[1] by this means, however, only the surface is exposed to inspection. As regards the general condition of the cerebral circulation, some idea may be obtained by manometric examination of the pressure within the skull, but this procedure, as far as is known, has never been applied to pharmacological studies.[2]

V.

INVESTIGATION AND EXPLANATION OF THE ANATOMICAL ALTERATIONS PRODUCED BY POISONS.

THE anatomical changes produced by the action of poisons can be studied only in the most limited degree during life, as in inflammation at the point of absorption, when visible to the eye, or increase or decrease of fatty tissue, etc. Most anatomical changes can only be recognized after death, and it is, therefore, advisable, in order to obtain a complete picture of the morbid processes, to kill the animals experimented on in different stages of the poisoning.

No general rules can be given for the investigation of these points, since they will depend upon the nature of the pathological process. An attempt, however, should invariably be made to determine whether the changes are due to direct action on the tissue concerned, or whether they are secondary results from some functional disturbance. In general, the latter state of affairs will be found to exist, except of course in cases in which the results are evidently attributable to inflammation of the absorbing surface.

---

[1] The procedure is described by Krause, in Anat. des Kaninchens, p. 46.

[2] Various methods of investigation of the cerebral circulation are given in the Medical Lancet, Oct. 1850, Moleschott's Untersuch., iii., Virchow's Archiv, xxxvii. 519, and Monographs by Jolly, Würzburg, 1871, Althann, Dorpat, 1871, Pagenstecher, Heidelberg, 1871.

A large number of poisons, among which are the nerve poisons which probably produce some obscure chemical change in the nerve elements, leave no detectable post-mortem trace of their action, thus proving that the most profound functional disturbance may exist without any palpable tissue change.

In the examination of post-mortem appearances, account must be taken of the changes which can be attributed to the act of dying, and it is not sufficient merely to separate the normal post-mortem processes of coagulation in blood and muscle and the changes of decomposition, from the effects due to the action of the drug. In most forms of death, a large number of changes, in no way characteristic of the special cause of death, are produced by the stoppage of circulation and respiration. Among these, the ordinary asphyxic appearances are the most usual, since nearly every form of death occurs under the symptoms of suffocation. It is further to be noted that the characteristic changes produced by the poison, where such exist, may be modified or removed during the act of dying. Moreover, no reliable conception can be obtained as to the state of the cerebral circulation, or the amount of fluid in the brain, from the post-mortem changes, since such conditions are sure to be disturbed in the death struggle, and the uncertainty will be the greater the longer the act of dying is prolonged. When, therefore, it is desired to examine into these points, the animal should be killed by puncture of the medulla, so as to prevent the signs of dyspnœa or convulsions being confounded with those directly due to the poison.

Alterations due directly to the action of the poison can, with any degree of certainty, be only detected at the point of absorption of the poison; particularly when the poison is of a diffusive character, and so apt to penetrate deeply into the tissues without following the vessels. Such changes are usually of a corrosive character; that is, chemical destruction of the tissue elements, with its consequent inflammatory changes.

[Degenerative changes and changes in the histological elements at the point of application will also fall under the heading of local action. Thus Eröss found that injections of oil of mustard into muscles caused fatty degeneration and disappearance of their transverse striæ.]

These alterations must, from their very nature, be restricted to the points with which the poison comes in direct contact, since after absorption by the bloodvessels the poison either undergoes chemical change, or is so diluted as to be unable to produce its corrosive effects. The action of poisons administered by the alimentary canal may, of course, be spread over a very extended surface. Usually, however, the corrosive action of a poison is restricted to the point of application, and to the depth to which, in a concentrated form, it is capable of diffusing.

When the absorbing surface has thin walls, as is the case in the stomach or intestines, the irritant action may extend through diffusion, by continuity of surface, to neighboring organs, such as the liver, spleen, or diaphragm. If the action leads to perforation, the poison may gain access to the body cavities, and thus come in contact with greatly increased surfaces.

In addition to this irritant action and its sequellæ (such as hyperæmia, catarrh, swelling, suppuration, infiltration, degeneration, and scarring), but few specific toxic anatomical changes can be detected. When such exist, they are generally to be found in the glands and muscles (fatty degeneration), and skin (exanthemata), and are only produced when the action of the poison has not been rapidly fatal. The connecting links between the direct action of the poison and these secondary results are hidden in the greatest obscurity. It can only be said that they are due to changes in nutrition without any explanation being possible. In many cases the locality of these changes appears to depend upon the course of the poison in the system; thus kidney changes may be produced by drugs eliminated through the urine, so, perhaps, showing that they are due to direct anatomical action of the poison.

VI.

INVESTIGATION OF CHEMICAL CHANGES PRODUCED BY POISONS.

OF the chemical processes resulting from the administration of drugs (see page 80), only those will be here
alluded to which may serve to add to our knowledge
of the mode of action of the drug. Changes in the
composition of the poison will not here come under
consideration. These investigations, so far as they do
not concern alteration of the secretions, can only be
undertaken post mortem, and here, even more than in
the study of anatomical alterations, great care must be
taken in excluding the results of post-mortem changes;
consequently, the examinations should be made immediately after death. Since the normal chemical constituents of different organs are but imperfectly known,
and the actions of poisons probably involve obscure .
chemical changes, but little can be here said; especially
as the constituents, of which our knowledge is at all complete, are either products of tissue change, already des-
tined for elimination, or decomposition products of the
unstable tissue elements, produced in post-mortem changes
or in the chemical manipulations required in the analysis.

In the case of nerve and muscle poisons it cannot be
stated whether their action is due to chemical change
or not. In the case of the direct action of poisons on the
blood, as far as combinations with the gases of the blood
are concerned, our information is more definite.

The study of toxic disturbances of nutrition often produced by poisons, with the exception of estimation of
sugar and glycogen, when post-mortem changes should be
prevented by boiling water, are probably better carried on
indirectly through examination of the excretions, rather
than in direct analysis of the different organs.

The greatest number of investigations as to the post-
mortem presence of poisons in the body have been undertaken for medico-legal purposes, a class of study which
does not fall within the province of this work.

APPENDIX.

1. DOSES, IMMUNITIES, FORM, AND SOLVENTS OF POI
SONS.—To obtain a complete idea as to the action of a
poison, it must be administered in doses varying from
the smallest active quantity to maximum doses; that is,
until a point has been reached when increase in the dose
causes no increase in the character or intensity of the
resulting symptoms. This is not only necessary on practical grounds (since to be of therapeutic value, it must be
known what quantities are fatal and what is the smallest
amount that will produce any effect), but more especially
because very often the effects of a poison will be found
to vary with the quantity used. This graduation of dose
must be established for each class of animal experimented
on, since it will often be found that corresponding doses
will produce different results in different classes of animals. Many poisons are entirely inert on certain species,
while very virulent on others; but occasionally it will
be found that this immunity is only apparent, the result
being due merely to a difference in the dose required.
When such immunities are detected it is an interesting
question to attempt to explain their cause. Before all,
an apparent immunity must be separated from one which
is actual; apparent immunities exist when, although the
fundamental action of the poison is exerted, it causes only
such functional disturbances as do not produce any evident
effect on the organism; such as the apparent immunity
of frogs to carbonic oxide. In actual immunities the action of the poison is not exerted, or only after the administration of disproportionately large doses. The cause of

such immunities cannot be positively stated, though the following may be urged as possible explanations : a, the chemical conditions necessary for the action of the poison do not exist in the animal experimented on; an example of this is seen in the immunity enjoyed by insects for carbonic acid, since their respiration is not carried on with hæmoglobin ; b, the nutritive changes occurring in the animal may be of such a character as to cause such an unusually rapid excretion or decomposition of the poison as not to allow the accumulation in the blood of a quantity sufficient to produce toxic action. Such a process may be established when the prevention of excretion, as by ligation of the ureters, renders a much smaller dose active. Actual immunities may be determined by the detection of the poison in the animal tissues ; thus, the above explanation will not apply to the immunity of rabbits to belladonna, for their flesh, after they themselves have received unharmed large quantities of belladonna, will produce symptoms of atropia poisoning when consumed by man or other animals. Under no circumstances can any explanation be attempted unless all the modes of action of the poison are thoroughly understood. With the exception of cases in which special immunities exist, the general law holds good that the dose required to produce the general action of a poison is in accordance with the size of the animal ; this is readily understood when it is remembered that a certain percentage of poison in the blood is required before the effects of the poison are manifested, and of course as the amount of blood contained in a small animal is less than that of a larger one, a smaller dose is required to produce proportionate effects. This rule has, however, many exceptions.

In many cases it is necessary to repeat the doses at intervals in order to prolong the period of poisoning, and permit extended observation of any particular stage ; often, however, this will be impossible, as animals will frequently lose their susceptibility to repeated doses of a poison. Such reduced susceptibility occurs with nicotin, and, indeed, remains after the cessation of the administration of the drug; such conditions may be described as acquired immunities, or the animal or person is said to become habituated to the poison: its explanation is one of the most difficult problems of pharmacology.

The repeated administration of drugs is further necessary from another point of view, since many poisons only produce certain effects after repeated administration: ordinarily under such circumstances the organism undergoes such changes that a single dose is not capable of producing the characteristic effects of the drug. Thus a number of poisons require prolonged contact, or that of their products, with the special organs on which they act; consequently a single dose is not capable, on account of rapid elimination, of producing its typical action. The effect cannot even be produced by increasing the dose of the poison; because, on the one hand, it may cause death by some other mode of action, or, on the other hand, there is a limit to the quantity of poison capable of absorption, especially in the more insoluble poisons in which the amount administered is of little consequence, as the excess over and above that which can be absorbed is carried off by the feces, when given by the stomach, or by suppuration when given subcutaneously. In such cases the repeated administration of the poison is indispensable for the production of the characteristic action.

Such cases are designated by the expression chronic poisoning, an unfortunate term, since chronic .poisoning is not separated from the acute form by the time required for its development, or the duration of its existence, but by the fact that acute poisoning is caused by a single dose, chronic poisoning by doses repeated at intervals. (1) To this class of chronic poisonings belong the cases of so-called cumulative action; that is, effects which only appear after the repeated administration of separate doses, even though they be small, and which cannot be caused by the administration of a single dose, even though it be a large one. The cause can only lie in the explanation given above.

Another case in which the characteristic action may only be produced by repeated small doses is when larger doses, by some special action, such as vomiting, or defecation, either prevent absorption, or cause such rapid elimination, that either no action, or a modified one, occurs.

The physical form in which the poison is administered is of great influence on its mode and conditions of action. It is only when in solution that a poison can enter the blood by absorption, and the rapidity of absorption depends largely on the character of the solvent; and upon the rapidity of absorption will depend the character or even the existence of the general action. Poisons, introduced in the solid form, often find solvents in the different fluids of the body, especially in the water of the tissue juices and secretions, which will permit this absorption. These natural solvents should not, however, be relied upon, but the drug be invariably administered in solution, since the abstraction of water from the tissues may set up inflammatory changes which might complicate the result, or the solvent might not be on hand in sufficient quantity to dissolve the amount of drug administered ; thus, for example, it is possible that in acute phosphorus poisoning, the phosphorus is absorbed without producing local inflammatory changes when fat is found at the same time in the digestive canal.

In the choice of a solvent, one must be selected that is itself absorbable and indifferent (inactive), and the solution should not be too concentrated, as concentrated solutions, like the solid body, tend to produce corrosive effects.

2. METHODS OF PRODUCING NARCOSIS.—For a number of special pharmacological studies it is necessary, before giving the poison under study, to give some drug which will destroy certain functions which it is desired to

eliminate. Ordinarily it is desired to maintain the animal in a passive condition, in which no voluntary motions will be made, and curare, therefore, is the drug
most frequently used, either to produce absolute motion-
lessness in complicated experiments, as in blood-
pressure experiments, or to eliminate, through paralysis
of the motor nerve endings, the possible effect of the
drug on these organs. In all cases of curare poisoning
in warm-blooded animals, artificial respiration must be
maintained. Curare must not be used in the study of·
drugs which, *a*, act themselves on the motor apparatus,
or, *b*, which produce diabetes, since the latter is also
caused by curare, or, *c*, it must be used only in very
small doses when the drug under study itself acts on the
vaso-motor nerves.

Whenever any drug is used for these purposes, all its
actions must be thoroughly understood, and must not be
incompatible with the drug under study, unless the line
of incompatibility can be sharply drawn, and may itself
serve such purposes as above alluded to.

Chloral is often employed to destroy pain and keep
the animal motionless, and does not require the main-
tenance of artificial respiration; it is especially suited
for rabbits, which, as a rule, do not stand narcotics well.
Chloral, however, cannot be used in the study of the
action of poisons on the heart, vessels, or pupil.

Morphia, or laudanum, may be used for the same purposes, and with the same limitations; both are well suited
for dogs.

These drugs are not, however, as yet thoroughly
enough understood to allow of certainty that they exert no
antagonism on the action of the special poison which may
be the subject of study; they should therefore only be
used in special cases, and control experiments should
always be made without the use of any narcotic in order
to determine the general action of any drug.

17

Antagonism of Drugs.

[No study of the action of a drug can be considered complete unless some attempt has been made to discover its physiological antidote. The chemical antidotes of a drug, when such exist, are usually readily determined by its known chemical incompatibilities. The practical value of such knowledge is, however, restricted in cases of actual poisoning to the time during which the poison remains within the alimentary canal; thus alkalies can only be useful immediately after the ingestion of acids; iron hydrated sesquioxide, immediately after the administration of arsenic. By physiological antagonism, however, as expressed by Bartholow,[1] is meant a balance of opposed actions on particular organs or tissues. This antagonism, or opposition of actions, may extend throughout the whole range of effects of two different drugs, or it may be limited to a few points; and, indeed, some of the most valuable instances of antagonism are thus limited, and there are few, if any, examples in which the opposition of actions is universal. In the search, therefore, for a physiological antagonist, a drug should be first selected in which the points of contrast in physiological action are most marked. Thus, when it has been determined that the poison under study destroys life by paralysis of the respiratory centre, a marked stimulant of that centre should first be tested as to its possession of antidotal powers. A cardiac depressant should be antagonized with a cardiac stimulant, etc.

When a drug has been selected which offers the greatest number, or most pronounced, points of contrast to the poison under study, the first point, if not already settled, will be to determine the minimum fatal dose of the poison per pound weight of the animals experimented on; then the minimum fatal dose of the drug which it is proposed to test as an antidote. This having been

---

[1] Antagonism between Medicines, Cartwright Lectures, 1880.

accomplished, the minimum lethal dose of the poison is administered, to be followed by the administration within a few minutes of the corresponding dose of the antidote. Should the animal survive, the same dose of the poison should then be administered a few days later to test whether the dose originally given was sufficiently large to cause death. After a drug which possesses antidotal properties has been found, and the proper dose determined, the antidote and the poison may be given simultaneously. Experiments should also be made as to the time which may elapse and the poisoned animal still be saved by the antidote, and as to how much more than the minimum fatal dose of the poison may be given and the animal's life still be preserved by the antidote. A curious fact which has been over and over again demonstrated, and should be remembered in such investigations, is that when less than the minimum fatal doses of two poisons, which modify each other's action, are given simultaneously, death will often result.

After a successful antagonism has been proved, it will then be interesting to see in what manner the fatal effects are prevented. To that end, if the drug is a circulatory poison, for instance, paralyzing the vagi and vaso-motor centre, a blood-pressure experiment should be made, and when the characteristic effects have been produced, an appropriate dose of the antidote should be given and the effects on the circulation noted; whether the blood-pressure rises and the vagi and vaso-motor centre regain their irritability. If the action is on the heart or respiratory centre, experiments such as those already detailed under the heading of the circulation or respiration may be instituted.]

INDEX.

17*

CATALOGUE OF BOOKS

PUBLISHED BY

LEA BROTHERS & CO.

(LATE HENRY C. LEA'S SON & CO.)

The books in the annexed list will be sent by mail, post-paid, to any Post Office in the United States, on receipt of the printed prices. No risks of the mail, however, are assumed, either on money or books. Gentlemen will, therefore, in most cases, find it more convenient to deal with the nearest bookseller.

In response to a large number of inquiries for a finer binding than is usually placed on medical books, we now finish many of our standard publications in half Russia, using in the manufacture none but the best materials. To foster the growing taste, the prices have been fixed at so small an advance over the cost of leather binding as to bring it within the reach of all to possess a library attractive to the eye as well as to the mind.

Detailed catalogues furnished or sent free by mail on application.

LEA BROTHERS & CO.,

(10.9.)   Nos. 706 and 708 Sansom Street, Philadelphia.

Periodicals. 1889.

THE MEDICAL NEWS,

A WEEKLY JOURNAL OF MEDICAL SCIENCE,

Published every Saturday, containing 28-32 large double-columned quarto pages of reading matter in each number.

FIVE DOLLARS ($5) per annum, in advance.

MONTHLY PUBLICATION OF

THE AMERICAN JOURNAL OF THE MEDICAL SCIENCES.

EDITED BY I. MINIS HAYS, A.M., M.D.

With the issue of January, 1888, THE AMERICAN JOURNAL OF THE MEDICAL SCIENCES was changed from a quarterly to a monthly, being enlarged to contain 112 pages in each number. Simultaneously the price was

REDUCED TO FOUR DOLLARS PER ANNUM.

COMMUTATION RATE.

THE MEDICAL NEWS ($5)
THE AMERICAN JOURNAL OF MEDICAL SCIENCES ($4) } **$7.50 per annum, in advance.**

THE MEDICAL NEWS VISITING LIST.

This LIST, which is by far the most handsome and convenient now attainable, has been thoroughly revised for 1890. It contains 48 pages of useful data, including the latest trustworthy therapeutical novelties with their properties and doses, and 176 pages of ruled blanks for various memoranda, and it is furnished with flap, pocket, pencil, erasable tablet, and catheter scale. It is issued in three styles—Weekly (dated, for 30 patients), Monthly (undated, for 120 patients), and Perpetual (undated). Each in one volume, price, $1 25. Advance paying subscribers to either or both the above periodicals may obtain The Medical News Visiting List for 75 cents. Or Journal, News, Visiting List, and Year Book of Treatment (see p. 16) to one address, $8 50. Thumb-letter Index for quick use 25 cents extra.

ALLEN (HARRISON). A SYSTEM OF HUMAN ANATOMY. WITH AN INTRODUCTORY SECTION ON HISTOLOGY, by E. O. Shakespeare, M.D. Comprising 813 double-columned quarto pages, with 380 engravings on stone on 109 plates, and 241 woodcuts in the text. In six sections, each in a portfolio. Section I. (Histology), Section II. (Bones and Joints), Section III. (Muscles and Fasciæ), Section IV. (Arteries, Veins and Lymphatics), Section V. (Nervous System), Section VI. (Organs of Sense, of Digestion and Genito-Urinary Organs, Embryology, Development, Teratology, Post Mortem Examinations, General and Clinical Indexes). Price per section, $3 50. Also, bound in one volume, cloth, $23. *Sold by subscription only.*

AMERICAN SYSTEM OF DENTISTRY. In treatises by various authors. Edited by Wilbur F. Litch, M.D., D.D.S. In three very handsome super-royal octavo volumes, containing 3180 pages, with 2863 illustrations and 9 full-page plates. *Now ready* Per volume, cloth, $6; leather, $7; half Morocco, $8. *For sale by subscription only.* Apply to the publishers.

AMERICAN SYSTEMS OF GYNECOLOGY AND OBSTETRICS. In treatises by the most eminent American specialists. Gynecology edited by Matthew D. Mann, A M., M.D., and Obstetrics edited by Barton C. Hirst, M.D. In four large octavo volumes comprising 3612 pages, with 1092 engravings, and 8 colored plates. Complete work *just ready.* Per volume, cloth, $5; leather, $6; half Russia, $7. *For sale by subscription only.* Prospectus free on application to publishers. ·

ASHHURST (JOHN, Jr.) THE PRINCIPLES AND PRACTICE OF SURGERY. FOR THE USE OF STUDENTS AND PRACTITIONERS. New (fifth) and revised edition. In one large and handsome octavo volume of 1148 pages, with 642 woodcuts. Cloth, $6; leather, $7. *Just ready.*

ASHWELL (SAMUEL). A PRACTICAL TREATISE ON THE DISEASES OF WOMEN. Third edition. 520 pages. Cloth, $3 50.

A SYSTEM OF PRACTICAL MEDICINE BY AMERICAN AUTHORS. Edited by William Pepper, M.D., LL.D. In five large octavo volumes, containing 5573 pages and 198 illustrations. Price per volume, cloth, $5 00; leather, $6 00; half Russia, $7 00. *Sold by subscription only.* Address the publishers.

ATTFIELD (JOHN). CHEMISTRY; GENERAL, MEDICAL AND PHARMACEUTICAL. Twelfth edition, specially revised by the Author for America. In one handsome 12mo. volume of 782 pages, with 88 illustrations. Cloth, $2 75; leather, $3 25. *Just ready.*

BALL (CHARLES B.) DISEASES OF THE RECTUM AND ANUS. In one 12mo. vol. of 417 pages, with 54 illus. and 4 colored plates. Cloth, $2 25. See *Series of Clinical Manuals,* p. 13.

BARKER (FORDYCE). OBSTETRICAL AND CLINICAL ESSAYS. In one handsome 12mo. volume of about 300 pages. *Preparing.*

BARLOW (GEORGE H.) A MANUAL OF THE PRACTICE OF MEDICINE. In one 8vo. volume of 603 pages. Cloth, $2 50.

BARNES (ROBERT). A PRACTICAL TREATISE ON THE DISEASES OF WOMEN. Third American from 3d English edition. In one 8vo. vol. of about 800 pages, with about 200 illus. *Preparing.*

BARNES (ROBERT and FANCOURT). A SYSTEM OF OBSTETRIC MEDICINE AND SURGERY, THEORETICAL AND CLINICAL. The Section on Embryology by Prof. Milnes Marshall. In one large octavo volume of 872 pages, with 231 illustrations. Cloth, $5; leather, $6.

BARTHOLOW (ROBERTS). MEDICAL ELECTRICITY. A PRAC-
TICAL TREATISE ON THE APPLICATIONS OF ELECTRICITY
TO MEDICINE AND SURGERY. Third edition. In one 8vo
vol. of 308 pages, with 110 illustrations. Cloth, $2 50.
—— NEW REMEDIES OF INDIGENOUS SOURCE, THEIR PHY-
SIOLOGICAL ACTIONS AND THERAPEUTICAL USES. In
one octavo volume of about 300 pages. *Preparing.*

BASHAM (W. R.) RENAL DISEASES; A CLINICAL GUIDE TO
THEIR DIAGNOSIS AND TREATMENT. In one 12m · volume
of 304 pages, with illustrations. Cloth, $2 00

BELLAMY (EDWARD). A MANUAL OF SURGICAL ANATOMY.
In one 12mo. vol. of 300 pges, with 50 illustrations. Cloth, $2 25.

BELL (F. JEFFREY). COMPARATIVE PHYSIOLOGY AND ANAT-
OMY. In one 12mo. volume of 561 pages, with 229 woodcuts.
Cloth, $2. See *Students' Series of Manuals*, p. 14.

BERRY (GEORGE A.) DISEASES OF THE EYE; A PRACTICAL
TREATISE FOR STUDENTS OF OPHTHALMOLOGY. Very
handsome octavo, 685 pages, with 144 original illustrations in the
text, of which 62 are exquisitely colored. Cloth, $7 50. *Just
ready.*

BILLINGS (JOHN S.) THE NATIONAL MEDICAL DICTIONARY.
Including in one alphabet English, French, German, Italian, and
Latin Technical Terms used in Medicine and the Collateral Sciences
For sale by subscription only. Specimen pages on application to
publishers. *In Press.*

BLOXAM (C. L.) CHEMISTRY, INORGANIC AND ORGANIC.
With Experiments. New American from the fifth London edition.
In one handsome octavo volume of 727 pages, with 292 illustra-
tions. Cloth, $2; leather, $3.

BRISTOWE (JOHN SYER). A TREATISE ON THE PRACTICE OF
MEDICINE. Second American edition, revised by the Author.
Edited with additions by James H. Hutchinson, M.D. In one
8vo. vol. of 1085 pp. Cloth, $5; leather, $6.

BROADBENT. (W. H.) THE PULSE. *Preparing.* See *Series of
Clinical Manuals*, p. 13.

BROWNE (EDGAR A.) HOW TO USE THE OPHTHALMOSCOPE.
Elementary instruction in Ophthalmoscopy for the Use of Students.
In one small 12mo. volume of 116 pages, with 35 illust. Cloth, $1.

BRUCE (J. MITCHELL). MATERIA MEDICA AND THERA-
PEUTICS. Fourth edition. In one 12mo. volume of 591 pages.
Cloth, $1 50. See *Students' Series of Manuals*, p. 14.

BRUNTON (T. LAUDER). A MANUAL OF PHARMACOLOGY,
THERAPEUTICS AND MATERIA MEDICA; including the
Pharmacy, the Physiological Action and the Therapeutical Uses of
Drugs. New (third and revised) edition, in one octavo volume of
1305 pages, with 230 illustrations. Cloth, $5 50; leather, $6 50.

BRYANT (THOMAS). THE PRACTICE OF SURGERY. Fourth
American from the fourth English edition. In one imperial octavo
volume of 1040 pages, with 727 illustrations. Cloth, $6 50;
leather, $7 50.

BUMSTEAD (F. J.) and TAYLOR (R. W.) THE PATHOLOGY AND
TREATMENT OF VENEREAL DISEASES. New edition. See
Taylor on Venereal Diseases.

BURNETT (CHARLES H.) THE EAR: ITS ANATOMY, PHYSI-
OLOGY AND DISEASES. A Practical Treatise for the Use of
Students and Practitioners. Second edition. In one 8vo. vol. of
580 pp., with 107 illus. Cloth, 4; leather, $5.

BUTLIN, (HENRY T.) DISEASES OF THE TONGUE. In one pocket-size 12mo. vol. of 456 pp., with 8 col. plates and 3 woodcuts. Limp cloth, $3 50. See *Series of Clinical Manuals*, p. 13.

CARPENTER WM. B.) PRIZE ESSAY ON THE USE OF ALCOHOLIC LIQUORS IN HEALTH AND DISEASE. New Edition, with a Preface by D. F. Condie, M.D. One 12mo. volume of 178 pages. Cloth, 60 cents.

—— PRINCIPLES OF HUMAN PHYSIOLOGY. A new American, from the eighth English edition. In one large 8vo. volume.

CARTER (R. BRUDENELL) AND FROST (W. ADAMS). OPHTHALMIC SURGERY. In one pocket-size 12mo. volume of 559 pages, with 91 engravings and one plate. Cloth, $2 25. See *Series of Clinical Manuals*, p. 13.

CHAMBERS (T. K.) A MANUAL OF DIET IN HEALTH AND DISEASE. In one handsome 8vo. vol. of 302 pages. Cloth, $2 75.

CHAPMAN (HENRY C). A TREATISE ON HUMAN PHYSIOLOGY. In one octavo volume of 925 pages, with 605 illustrations. Cloth, $5 50; leather, $6 50.

CHARLES (T. CRANSTOUN). THE ELEMENTS OF PHYSIOLOGICAL AND PATHOLOGICAL CHEMISTRY. In one handsome octavo volume of 451 pages, with 38 woodcuts and one colored plate. Cloth, 3 50.

CHURCHILL (FLEETWOOD). ESSAYS ON THE PUERPERAL FEVER. In one octavo volume of 464 pages. Cloth, $2 50.

CLARKE (W. B.) AND LOCKWOOD (C. B.) THE DISSECTOR'S MANUAL. In one 12mo. volume of 396 pages,'with 49 illustrations. Cloth, $1 50. See *Students' Series of Manuals*, p. 14.

CLASSEN'S QUANTITATIVE ANALYSIS. Translated by Edgar F. Smith, Ph.D. In one 12mo. vol. of 324 pp., with 36 illus. Cloth, $2 00.

CLELAND (JOHN). A DIRECTORY FOR THE DISSECTION OF THE HUMAN BODY. In one 12mo. vol. of 178 pp. Cloth, $1 25.

CLOUSTON (THOMAS S.) CLINICAL LECTURES ON MENTAL DISEASES. With an Abstract of Laws of U. S. on Custody of the Insane, by C. F. Folsom, M.D. In one handsome octavo vol. of 541 pages, illustrated with woodcuts and 8 lithographic plates. Cloth, $4 00. Dr. Folsom's *Abstract* is also furnished separately in one octavo volume of 108 pages. Cloth, $1 50.

CLOWES (FRANK). AN ELEMENTARY TREATISE ON PRACTICAL CHEMISTRY AND QUALITATIVE INORGANIC ANALYSIS. New American from the fourth English edition. In one handsome 12mo. volume of 387 pages, with 55 illustrations. Cloth, $2 50.

COATS (JOSEPH). A TREATISE ON PATHOLOGY. In one vol. of 829 pp., with 339 engravings. Cloth, $5 50; leather, $6 50.

COHEN (J. SOLIS). DISEASES OF THE THROAT AND NASAL PASSAGES. Third edition, thoroughly revised. In one handsome octavo volume. *Preparing*.

COLEMAN (ALFRED). A MANUAL OF DENTAL SURGERY AND PATHOLOGY. With Notes and Additions to adapt it to American Practice. By Thos. C. Stellwagen, M.A., M.D., D.D.S. In one handsome 8vo. vol. of 412 pp., with 331 illus. Cloth, $3 25.

CONDIE (D. FRANCIS). A PRACTICAL TREATISE ON THE DIS-
EASES OF CHILDREN. Sixth edition, revised and enlarged. In
one large 8vo. vol. of 719 pages. Cloth, $5 25; leather, $6 25.

COOPER (B. B.) LECTURES ON THE PRINCIPLES AND PRACTICE
OF SURGERY. In one large 8vo. vol. of 767 pages. Cloth, $2 00.

CORNIL (V.) SYPHILIS: ITS MORBID ANATOMY, DIAGNOSIS
AND TREATMENT. Translated, with notes and additions, by J.
Henry C. Simes, M.D , and J. William White, M.D. In one 8vo.
volume of 461 pages, with 84 illustrations. Cloth, $3 75.

CULLERIER (A.) AN ATLAS OF VENEREAL DISEASES. Translated and edited by FREEMAN J. BUMSTEAD, M.D., LL.D. A large
quarto volume of 328 pages, with 26 plates containing about 150
figures, beautifully colored, many of them life-size. Cloth, $17.

DALTON (JOHN C.) DOCTRINES OF THE CIRCULATION OF
THE BLOOD. In one handsome 12mo. vol. of 293 pp. Cloth, $2.

—— A TREATISE ON HUMAN PHYSIOLOGY. Seventh edition,
thoroughly revised, and greatly improved. In one very handsome
8vo. vol. of 722 pages, with 252 illustrations. Cloth, $5; lea-
ther, $6.

DANA (JAMES D.) THE STRUCTURE AND CLASSIFICATION OF
ZOOPHYTES. With illust. on wood. In one imp. 4to. vol. Cl., $4.

DAVENPORT (F. H.) DISEASES OF WOMEN. A Manual of Non-
Surgical Gynæcology. For the use of Students and General Practitioners In one handsome 12mo. volume of 306 pages with 105
illustrations. Cloth, $1 50. Just ready.

DAVIS (F. H.) LECTURES ON CLINICAL MEDICINE. Second
edition In one 12mo. volume of 287 pages. Cloth, $1 75.

DE LA BECHE'S GEOLOGICAL OBSERVER. In one large 8vo. vo
of 700 pages, with 300 illustrations. Cloth, $4.

DRAPER (JOHN C.) MEDICAL PHYSICS. A Text-book for Students and Practitioners of Medicine. In one handsome octavo volume of 734 pages, with 376 illustrations. Cloth, $4.

DRUITT (ROBERT). THE PRINCIPLES AND PRACTICE OF
MODERN SURGERY. A new American from the 12th London
edition, edited by Stanley Boyd, F.R.C.S. In one large octavo
volume of 965 pages, with 373 illustrations. Cloth, $4; leather, $5.

DUNCAN (J. MATTHEWS). CLINICAL LECTURES ON THE DIS-
EASES OF WOMEN. Delivered in St. Bartholomew's Hospital.
In one octavo volume of 175 pages. Cloth, $1 50.

DUNGLISON (ROBLEY). MEDICAL LEXICON; A Dictionary of
Medical Science. Containing a concise explanation of the various
subjects and terms of Anatomy, Physiology, Pathology, Hygiene,
Therapeutics, Pharmacology, Pharmacy, Surgery, Obstetrics, Medical Jurisprudence and Dentistry; notices of Climate and of Mineral
Waters; Formulæ for Officinal, Empirical and Dietetic Preparations;
with the accentuation and Etymology of the Terms, and the French
and other Synonymes. Edited by R. J. Dunglison, M.D. In one
very large royal 8vo. vol. of 1139 pages. Cloth, $6 50; leather,
$7 50; half Russia, $8.

EDES' TEXT-BOOK OF THERAPEUTICS AND MATERIA MEDICA.
In one 8vo. volume of 544 pages. Cloth, $3 50; leather, $4 50.

EDIS (ARTHUR W.) DISEASES OF WOMEN. A Manual for Students and Practitioners. In one handsome 8vo. vol. of 576 pp.,
with 148 illustrations. Cloth, $3; leather, $4.

ELLIS (GEORGE VINER). DEMONSTRATIONS IN ANATOMY.
Being a Guide to the Knowledge of the Human Body by Dissection.
From the eighth and revised English edition. In one octavo vol.
of 716 pages, with 249 illustrations. Cloth. $4 25; leather, $5 25.

EMET (THOMAS ADDIS). THE PRINCIPLES AND PRACTICE
OF GYNÆCOLOGY, for the use of Students and Practitioners.
Third edition, enlarged and revised. In one large 8vo. volume of
880 pages, with 150 original illustrations. Cloth, $5; leather, $6.

ERICHSEN (JOHN E.) THE SCIENCE AND ART OF SURGERY.
A new American, from the eighth enlarged and revised London
edition. In two large octavo volumes containing 2316 pages, with
984 illus. Cloth, $9; leather, $11.

FARQUHARSON (ROBERT). A GUIDE TO THERAPEUTICS.
Fourth American from Fourth English edition, revised by Frank
Woodbury, M D. In one 12mo. volume of 581 pages. Cloth, $2 50.

FENWICK (SAMUEL). THE STUDENTS' GUIDE TO MEDICAL
DIAGNOSIS. From the third revised and enlarged London edition. In one royal 12mo. volume of 328 pages. Cloth, $2 25.

FINLAYSON (JAMES). CLINICAL DIAGNOSIS. A Handbook for
Students and Practitioners of Medicine. Second edition. In one
12mo. volume of 682 pages, with 158 illustrations. Cloth, $2 50.

FLINT (AUSTIN). A TREATISE ON THE PRINCIPLES AND
PRACTICE OF MEDICINE. Sixth edition, thoroughly revised
and largely rewritten by the Author, assisted by William H. Welch,
M.D , and Austin Flint, Jr., M.D. In one large 8vo. volume of
1160 pages, with illustrations. Cloth, $5 50; leather, $6 50.

—— A MANUAL OF AUSCULTATION AND PERCUSSION; of the
Physical Diagnosis of Diseases of the Lungs and Heart, and of Thoracic Aneurism. Fourth edition, revised and enlarged. In one
handsome 12mo. volume of 240 pages. Cloth, $1 75.

—— A PRACTICAL TREATISE ON THE DIAGNOSIS AND TREAT-
MENT OF DISEASES OF THE HEART. Second edition, enlarged.
In one octavo volume of 550 pages. Cloth, $4 00.

—— A PRACTICAL TREATISE ON THE PHYSICAL EXPLORA-
TION OF THE CHEST, AND THE DIAGNOSIS OF DISEASES
AFFECTING THE RESPIRATORY ORGANS. Second and revised
edition. In one octavo volume of 591 pages. Cloth, $4 50.

—— MEDICAL ESSAYS. In one 12mo. vol., pp. 210. Cloth, $1 38.

—— ON PHTHISIS: ITS MORBID ANATOMY, ETIOLOGY,
ETC. A series of Clinical Lectures. In one 8vo. volume of 442
pages. Cloth, $3 50.

FOLSOM (C. F.) An Abstract of Statutes of U. S. on Custody of the
Insane. In one 8vo. vol. of 108 pp. Cloth, $1 50. Also bound
with *Clouston on Insanity*.

FOSTER (MICHAEL). A TEXT-BOOK OF PHYSIOLOGY. Fourth American from the fifth English edition, edited by E. T. REICHERT, M.D. In one large 12mo. vol. of about 925 pages, with about 300 illustrations. *Preparing.*

FOTHERGILL'S PRACTITIONER'S HANDBOOK OF TREATMENT. New (third) edition. In one handsome octavo volume of 664 pages. Cloth, $3 75; leather, $4 75.

FOWNES (GEORGE). A MANUAL OF ELEMENTARY CHEMISTRY (INORGANIC AND ORGANIC). New edition. Embodying Watts' *Physical and Inorganic Chemistry.* In one royal 12mo. vol. of 1061 pages, with 168 illus., and one colored plate. Cloth, $2 75; leather, $3 25.

FOX (TILBURY) and T. COLCOTT. EPITOME OF SKIN DIS- EASES, with Formulæ. For Students and Practitioners. Third Am. edition, revised by T. C. Fox. In one small 12mo. volume of 238 pages. Cloth, $1 25.

FRANKLAND (E) and JAPP (F. R.) INORGANIC CHEMISTRY. In one handsome octavo vol. of 677 pages, with 51 engravings and 2 plates. Cloth, $3 75; leather, $4 75.

FULLER (HENRY). ON DISEASES OF THE LUNGS AND AIR PASSAGES. Their Pathology, Physical Diagnosis, Symptoms and Treatment. From 2d Eng. ed In 1 8vo. vol., pp. 475. Cloth, $3 50.

GIBNEY (V. P.) ORTHOPÆDIC SURGERY. For the use of Prac- titioners and Students. In one 8vo. vol. profusely illus. *Prepg.*

GIBSON'S INSTITUTES AND PRACTICE OF SURGERY. In two octavo volumes of 965 pages, with 34 plates. Leather, $6 50.

GLUGE (GOTTLIEB). ATLAS OF PATHOLOGICAL HISTOLOGY. Translated by Joseph Leidy, M.D., Professor of Anatomy in the University of Pennsylvania, &c. In one imperial quarto volume, with 320 copperplate figures, plain and colored. Cloth, $4

GOULD (A. PEARCE). SURGICAL DIAGNOSIS. In one 12mo. vol. of 589 pages. Cloth, $2. See *Students' Series of Manuals*, p. 14.

GRAY (HENRY). ANATOMY, DESCRIPTIVE AND SURGICAL. Edited by T. Pickering Pick, F.R.C.S. A new American, from the eleventh English edition, thoroughly revised, with additions, by W. W. Keen, M.D. To which is added Holden's "Landmarks, Medical and Surgical." In one imperial octavo volume of 1098 pages, with 685 large and elaborate engravings on wood. Cloth, $6; leather, $7; very handsome half Russia, raised bands, $7 50. The same edition is also issued with veins, arteries, and nerves distin- guished in colors. Price, cloth, $7 25; leather, $8 25; half Rus- sia, $8 75.

GRAY (LANDON CARTER). A PRACTICAL TREATISE ON THE DISEASES OF THE NERVOUS SYSTEM. In one handsome octavo volume of about 600 pages. *Preparing.*

GREEN (T. HENRY). AN INTRODUCTION TO PATHOLOGY AND MORBID ANATOMY. New (sixth) American, from the seventh London edition. In one handsome octavo volume of 540 pages, with 167 illustrations. Cloth, $2 75. *Just ready.*

GREENE (WILLIAM H.) A MANUAL OF MEDICAL CHEMISTRY. For the Use of Students. Based upon Bowman's Medical Chem- istry. In one 12mo. vol. of 310 pages, with 74 illus. Cloth, $1 75.

GRIFFITH (ROBERT E.) A UNIVERSAL FORMULARY, CON-TAINING THE METHODS OF PREPARING AND ADMINISTER-ING OFFICINAL AND OTHER MEDICINES. Third and enlarged edition. Edited by John M. Maisch, Phar.D. In one large 8vo. vol. of 775 pages, double columns. Cloth, $4 50; leather, $5 50.

GROSS (SAMUEL D.) A SYSTEM OF SURGERY, PATHOLOGICAL, DIAGNOSTIC, THERAPEUTIC AND OPERATIVE. Sixth edi-tion, thoroughly revised. In two imperial octavo volumes contain-ing 2382 pages, with 1623 illustrations. Strongly bound in leather, raised bands, $15.

—— A PRACTICAL TREATISE ON THE DISEASES, INJU-ries and Malformations of the Urinary Bladder, the Prostate Gland and the Urethra. Third edition, thoroughly revised and much condensed, by Samuel W. Gross, M.D. In one octavo volume of 574 pages, with 170 illus. Cloth, $4 50.

—— A PRACTICAL TREATISE ON FOREIGN BODIES IN THE AIR PASSAGES. In one 8vo. vol. of 468 pages. Cloth, $2 75.

GROSS (SAMUEL W.) A PRACTICAL TREATISE ON IMPO-TENCE, STERILITY, AND ALLIED DISORDERS OF THE MALE SEXUAL ORGANS. New (third) edition. In one hand-some octavo vol. of 163 pages, with 16 illustrations. Cloth, $1 50.

HABERSHON (S. O.) ON THE DISEASES OF THE ABDOMEN, AND OTHER PARTS OF THE ALIMENTARY CANAL. Second American, from the third English edition. In one handsome 8vo. volume of 554 pages, with illus. Cloth, $3 50.

HAMILTON (ALLAN McLANE). NERVOUS DISEASES, THEIR DESCRIPTION AND TREATMENT. Second and revised edition. In one octavo volume of 598 pages, with 72 illustrations. Cloth, $4.

HAMILTON (FRANK H.) A PRACTICAL TREATISE ON FRAC-TURES AND DISLOCATIONS. Seventh edition, thoroughly re-vised. In one handsome 8vo. vol. of 998 pages, with 352 illustra-tions. Cloth, $5 50; leather, $6 50.

HARTSHORNE (HENRY). ESSENTIALS OF THE PRINCIPLES AND PRACTICE OF MEDICINE. Fifth edition. In one 12mo. volume, 669 pages, with 144 illustrations. Cloth, $2 75; half bound, $3.

—— A HANDBOOK OF ANATOMY AND PHYSIOLOGY. In one 12mo. volume of 310 pages, with 220 illustrations. Cloth, $1 75.

—— A CONSPECTUS OF THE MEDICAL SCIENCES. Com-prising Manuals of Anatomy, Physiology, Chemistry, Materia Medica, Practice of Medicine, Surgery and Obstetrics. Second edition. In one royal 12mo. volume of 1028 pages, with 477 illus-trations. Cloth, $4 25; leather, $5 00.

HERMANN (L.) EXPERIMENTAL PHARMACOLOGY. A Hand-book of the Methods for Determining the Physiological Actions of Drugs. Translated by Robert Meade Smith, M.D. In one 12mo vol. of 199 pages, with 32 illustrations. Cloth, $1 50.

HILL (BERKELEY). SYPHILIS AND LOCAL CONTAGIOUS DIS-ORDERS In one 8vo. volume of 479 pages. Cloth, $3 25.

HILLIER (THOMAS). A HANDBOOK OF SKIN DISEASES. 2d ed. In one royal 12mo. vol. of 353 pp., with two plates. Cloth, $2 25.

HOBLYN (RICHARD D.) A DICTIONARY OF THE TERMS USED IN MEDICINE AND THE COLLATERAL SCIENCES. In one 12mo. vol. of 520 double-columned pp. Cloth, $1 50; leather, $2

HODGE (HUGH L.) ON DISEASES PECULIAR TO WOMEN, INCLUDING DISPLACEMENTS OF THE UTERUS. Second and revised edition. In one 8vo. volume of 519 pages. Cloth, $4 50.

———THE PRINCIPLES AND PRACTICE OF OBSTETRICS. In one large 4to. vol. of 542 double-columned pages, illustrated with large lithographic plates containing 159 figures from original photographs, and 110 woodcuts. Strongly bound in cloth, $14.

HOFFMANN (FREDERICK) AND POWER (FREDERICK B.) A MANUAL OF CHEMICAL ANALYSIS, as Applied to the Examination of Medicinal Chemicals and their Preparations. Third edition, entirely rewritten and much enlarged. In one handsome octavo volume of 621 pages, with 179 illustrations. Cloth, $4 25.

HOLDEN (LUTHER). LANDMARKS, MEDICAL AND SURGICAL. From the third English edition. With additions, by W. W. Keen, M.D. In one royal 12mo. vol. of 148 pp. Cloth, $1.

HOLLAND (SIR HENRY). MEDICAL NOTES AND REFLECTIONS. From 3d English ed. In one 8vo. vol. of 493 pp. Cloth, $3 50.

HOLMES (TIMOTHY). A SYSTEM OF SURGERY. With notes and additions by various American authors. Edited by John H. Packard, M.D. In three very handsome 8vo. vols. containing 3137 double-columned pages, with 979 woodcuts and 13 lithographic plates. Cloth, $18; leather, $21; very handsome half Russia, raised bands, $22 50. *For sale by subscription only.*

——— A TREATISE ON SURGERY. Its Principles and Practice. A new American from the fifth English edition. Edited by T. Pickering Pick, F.R.C.S. In one handsome octavo volume of 1008 pages, with 428 engravings. Cloth, $6; leather, $7. *Just ready.*

HORNER (WILLIAM E.) SPECIAL ANATOMY AND HISTOLOGY. Eighth edition, revised and modified. In two large 8vo. vols. of 1007 pages, containing 320 woodcuts. Cloth, $6.

HUDSON (A.) LECTURES ON THE STUDY OF FEVER. In one octavo volume of 308 pages. Cloth, $2 50.

HUTCHINSON (JONATHAN). SYPHILIS. In one pocket size 12mo. volume of 542 pages, with 8 chromo-lithographic plates. Cloth, $2 25. See *Series of Clinical Manuals*, p. 13.

HYDE (JAMES NEVINS). A PRACTICAL TREATISE ON DISEASES OF THE SKIN. New (second) edition. In one handsome octavo volume of 676 pages, with 85 engravings and 2 colored plates. Cloth, $4 50; leather, $5 50.

JONES (C. HANDFIELD). CLINICAL OBSERVATIONS ON FUNCTIONAL NERVOUS DISORDERS. Second American edition. In one octavo volume of 340 pages. Cloth, $3 25.

JULER (HENRY). A HANDBOOK OF OPHTHALMIC SCIENCE AND PRACTICE. In one 8vo. volume of 460 pages, with 125 woodcuts, 27 chromo-lithographic plates test types of Jaeger and Snellen and Holmgren's Color-blindness test. Cloth, $4 50; leather, $5 50.

KING (A. F. A.) A MANUAL OF OBSTETRICS. New (fourth) edition. In one 12mo. volume of 432 pages, with 141 illustrations. Cloth, $2 50. *Just ready*

KLEIN (E) ELEMENTS OF HISTOLOGY. Fourth edition. In one pocket-size 12mo. volume of 376 pages, with 194 engravings. Cloth, $1 75. *Just ready.* See *Students' Series of Manuals*, p. 14.

LANDIS (HENRY G) THE MANAGEMENT OF LABOR. In one handsome 12mo. volume of 329 pages, with 28 illus. Cloth, $1 75.

LA ROCHE (R.) YELLOW FEVER. In two 8vo. vols. of 1468 pages. Cloth, $7.

—— PNEUMONIA. In one 8vo. vol. of 490 pages. Cloth, $3.

LAURENCE (J. Z.) AND MOON (ROBERT C.) A HANDY-BOOK OF OPHTHALMIC SURGERY. Second edition, revised by Mr. Laurence. In one 8vo. vol. pp. 227, with 66 illus. Cloth, $2 75.

LAWSON (GEORGE). INJURIES OF THE EYE, ORBIT AND EYE-LIDS. From the last English edition. In one handsome octavo volume of 404 pages, with 92 illustrations. Cloth, $3 50.

LEA (HENRY C.) SUPERSTITION AND FORCE; ESSAYS ON THE WAGER OF LAW, THE WAGER OF BATTLE, THE ORDEAL AND TORTURE. Third edition, thoroughly revised and greatly enlarged. In one handsome royal 12mo. vol. pp. 552. Cloth, $2 50.

—— STUDIES IN CHURCH HISTORY. The Rise of the Temporal Power—Benefit of Clergy—Excommunication. New edition. In one handsome 12mo. vol. of 605 pp. Cloth, $2 50.

—— AN HISTORICAL SKETCH OF SACERDOTAL CELIBACY IN THE CHRISTIAN CHURCH. Second edition. In one handsome octavo volume of 684 pages. Cloth, $4 50.

LEE (HENRY) ON SYPHILIS. In one 8vo volume of 246 pages. Cloth, $2 25.

LEHMANN (C. G.) A MANUAL OF CHEMICAL PHYSIOLOGY. In one 8vo. vol. of 327 pages, with 41 woodcuts. Cloth, $2 25.

LEISHMAN (WILLIAM). A SYSTEM OF MIDWIFERY. Including the Diseases of Pregnancy and the Puerperal State. Third American, from the third English edition. With additions, by J. S. Parry, M.D. In one octavo volume of 740 pages, with 205 illustrations. Cloth, $4 50 ; leather, $5 50.

LUCAS (CLEMENT). DISEASES OF THE URETHRA. *Preparing.* See *Series of Clinical Manuals*, p. 13.

LUDLOW (J. L.) A MANUAL OF EXAMINATIONS UPON ANAT-OMY, PHYSIOLOGY, SURGERY, PRACTICE OF MEDICINE, OBSTETRICS, MATERIA MEDICA, CHEMISTRY, PHARMACY AND THERAPEUTICS. To which is added a Medical Formulary. Third edition. In one royal 12mo. volume of 816 pages, with 370 woodcuts. Cloth, $3 25 ; leather, $3 75.

LYONS (ROBERT D.) A TREATISE ON FEVER. In one octavo volume of 362 pages. Cloth, $2 25.

MAISCH (JOHN M.) A MANUAL OF ORGANIC MATERIA MED-ICA. New (third) edition. In one handsome 12mo. volume of 523 pages, with 257 beautiful illustrations. Cloth, $3.

MARSH (HOWARD). DISEASES OF THE JOINTS. In one 12mo. volume of 468 pages, with 64 illustrations and a colored plate. Cloth, $2. See *Series of Clinical Manuals*, p. 13.

MAY (C. H.) MANUAL OF THE DISEASES OF WOMEN. For the use of Students and Practitioners. In one 12mo. volume of 342 pages. Cloth, $1 75.

MEIGS (CHAS. D.) ON THE NATURE, SIGNS AND TREATMENT OF CHILDBED FEVER. In one 8vo. vol. of 346 pages. Cloth, $2.

MILLER (JAMES). PRINCIPLES OF SURGERY. Fourth American, from the third Edinburgh edition. In one large octavo volume of 688 pages, with 240 illustrations. Cloth, $3 75.

MILLER (JAMES). THE PRACTICE OF SURGERY. Fourth American, from the last Edinburgh edition. In one large octavo volume of 682 pages, with 364 illustrations. Cloth, $3 75.

MITCHELL (S. WEIR). LECTURES ON NERVOUS DISEASES, ESPECIALLY IN WOMEN. Second edition. In one 12mo. volume of 288 pages. Cloth, $1 75.

MORRIS (HENRY). SURGICAL DISEASES OF THE KIDNEY. 12mo., 554 pages, 40 woodcuts, and 6 colored plates. Cloth, $2 25. See *Series of Clinical Manuals*, p. 13.

MÜLLER (J.) PRINCIPLES OF PHYSICS AND METEOROLOGY. In one large 8vo. vol. of 623 pages, with 538 cuts. Cloth, $4 50.

NEILL (JOHN) AND SMITH (FRANCIS G.) A COMPENDIUM OF THE VARIOUS BRANCHES OF MEDICAL SCIENCE. In one handsome 12mo. volume of 974 pages, with 374 woodcuts. Cloth, $4; leather, raised bands, $4 75.

NETTLESHIP'S STUDENT'S GUIDE TO DISEASES OF THE EYE. New (third) edition. In one royal 12mo. volume of 479 pages, with 164 illustrations, test-types and formulæ. Cloth, $2.

NORRIS AND OLIVER ON THE EYE. In one 8vo. volume of about 500 pages, with illustrations. *Preparing.*

OWEN (EDMUND). SURGICAL DISEASES OF CHILDREN. 12mo., 525 pages, 85 woodcuts, and 4 colored plates. Cloth, $2. See *Series of Clinical Manuals*, p. 13.

PARRISH (EDWARD). A TREATISE ON PHARMACY. With many Formulæ and Prescriptions. Fifth edition, enlarged and thoroughly revised by Thomas S. Wiegand, Ph.G. In one octavo volume of 1093 pages, with 257 illustrations. Cloth; $5; leather, $6.

PARRY (JOHN S.) EXTRA-UTERINE PREGNANCY, ITS CLINICAL HISTORY, DIAGNOSIS, PROGNOSIS AND TREATMENT. In one octavo volume of 272 pages. Cloth, $2 50.

PARVIN (THEOPHILUS). THE SCIENCE AND ART OF OBSTETRICS. In one handsome 8vo. volume of 697 pages, with 214 engravings and a colored plate. Cloth, $4 25; leather, $5 25.

PAVY (F. W.) A TREATISE ON THE FUNCTION OF DIGESTION, ITS DISORDERS AND THEIR TREATMENT. From the second London edition. In one octavo volume of 238 pages. Cloth, $2.

PAYNE (JOSEPH FRANK). A MANUAL OF GENERAL PATHOLOGY. Designed as an Introduction to the Practice of Medicine. Handsome octavo volume of 524 pages with 153 engravings and 1 colored plate. Cloth, $3 50.

PEPPER (A. J.) FORENSIC MEDICINE. *In press.* See *Students' Series of Manuals,* p. 14.

—— SURGICAL PATHOLOGY. In one 12mo. volume of 5¹l pages, with 81 illus. Cloth, $2. See *Students' Series of Manuals,* p. 14.

PICK (T. PICKERING). FRACTURES AND DISLOCATIONS. In one 12mo. volume of 530 pages, with 93 illustrations, Cloth, $2. *See Series of Clinical Manuals,* p. 13.

PIRRIE (WILLIAM). THE PRINCIPLES AND PRACTICE OF SURGERY. In one handsome octavo volume of 780 pages, with 316 illustrations. Cloth, $3 75.

PLAYFAIR (W. S.) A TREATISE ON THE SCIENCE AND PRACTICE OF MIDWIFERY. New (fifth) American from the seventh English edition. Edited, with additions, by R. P. Harris, M.D. In one octavo volume of 664 pages, with 207 woodcuts and five plates. Cloth, $4; leather, $5. *Just ready.*

—— THE SYSTEMATIC TREATMENT OF NERVE PROSTRATION AND HYSTERIA. In one 12mo. vol. of 97 pages. Cloth, $1.

POLITZER (ADAM). A TEXT-BOOK OF THE EAR AND ITS DISEASES. Translated at the Author's request by James Patterson Cassells, M.D., F.F.P.S. In one handsome octavo volume of 800 pages, with 257 original illustrations. Cloth, $5 50.

POWER (HENRY). HUMAN PHYSIOLOGY. Second edition. In one 12mo. volume of 396 pages, with 47 illustrations. Cloth, $1 50. See *Students' Series of Manuals,* page 14.

PURDY ON BRIGHT'S DISEASE AND ALLIED AFFECTIONS OF THE KIDNEY. Octavo, 288 pp., with 18 handsome illus. Cloth, $2.

RALFE (CHARLES H.) CLINICAL CHEMISTRY. In one 12mo. volume of 314 pages, with 16 illustrations. Cloth, $1 50. See *Students' Series of Manuals,* page 14.

RAMSBOTHAM (FRANCIS H.) THE PRINCIPLES AND PRACTICE OF OBSTETRIC MEDICINE AND SURGERY. In one imperial octavo volume of 640 pages, with 64 plates, besides numerous woodcuts in the text. Strongly bound in leather, $7.

REMSEN (IRA). THE PRINCIPLES OF CHEMISTRY. New (third) edition, thoroughly revised, and much enlarged. In one 12mo. volume of 318 pages. Cloth, $2.

REYNOLDS (J. RUSSELL). A SYSTEM OF MEDICINE. Edited, with Notes and Additions, by HENRY HARTSHORNE, M.D. In three larga 8vo. vols., containing 3056 closely printed double-columned pages, with 317 illustrations. Per volume, cloth, $5; leather, $6; very handsome half Russia, $6 50. *For sale by subscription only*

RICHARDSON (BENJAMIN W.) PREVENTIVE MEDICINE. In one octa   ·· . of 729 pp. Clo , $4; leather, $5.

ROBERTS (JOHN B.) AND MORTON (THOS. S. K.) THE PRINCIPLES AND PRACTICE OF MODERN SURGERY. In one octavo volume of about 500 pages, fully illustrated. *Preparing.*

ROBERTS (JOHN B.) THE COMPEND OF ANATOMY. For use in the Dissecting Room and in preparing for Examinations. In one 16mo. volume of 196 pages. Limp cloth, 75 cents.

ROBERTS (WILLIAM). A PRACTICAL TREATISE ON URINARY AND RENAL DISEASES, INCLUDING URINARY DEPOSITS. Fourth American, from the fourth London edition. In one very handsome 8vo. vol. of 609 pages, with 81 illustrations. Cloth, $3 50.

ROBERTSON (J. McGREGOR). PHYSIOLOGICAL PHYSICS. In one 12mo. volume of 537 pages, with 219 illustrations. Cloth, $2 00. See *Students' Series of Manuals*, p. 14.

ROSS (JAMES). A HANDBOOK OF THE DISEASES OF THE NERVOUS SYSTEM. In one handsome octavo volume of 726 pages, with 184 illustrations. Cloth, $4 50; leather, $5 50.

SAVAGE (GEORGE H.) INSANITY AND ALLIED NEUROSES, PRACTICAL AND CLINICAL. In one 12mo. volume of 551 pages, with 18 typical illustrations. Cloth, $2 00. See *Series of Clinical Manuals*, p 13.

SCHÄFER (EDWARD A.) THE ESSENTIALS OF HISTOLOGY, DESCRIPTIVE AND PRACTICAL. For the use of Students. In one handsome octavo volume of 246 pages, with 281 illustrations. Cloth, $2 25.

SCHMITZ AND ZUMPT'S CLASSICAL SERIES. In royal 18mo. ADVANCED LATIN EXERCISES. Cloth, 60 cents; half bound, 70 cents.

SALLUST. Cloth, 60 cents; half bound, 70 cents.
NEPOS. Cloth, 60 cents; half bound, 70 cts.
VIRGIL. Cloth, 85 cents; half bound, $1.
CURTIUS. Cloth, 80 cents; half bound, 90 cents.

SCHOEDLER (FREDERICK) AND MEDLOCK (HENRY). WONDERS OF NATURE. An elementary introduction to the Sciences of Physics, Astronomy, Chemistry, Mineralogy, Geology, Botany, Zoology and Physiology. In one 8vo. vol., with 679 illus. Cloth, $3.

SCHREIBER (JOSEPH). A MANUAL OF TREATMENT BY MAS-SAGE AND METHODICAL MUSCLE EXERCISE. Translated by Walter Mendelson, M.D., of New York. In one handsome octavo volume of 274 pages, with 117 fine engravings. Cloth, $2 75.

SEILER (CARL). A HANDBOOK OF DIAGNOSIS AND TREAT-MENT OF DISEASES OF THE THROAT AND NASAL CAV-ITIES. New (3d) edition. In one very handsome 12mo. volume of 373 pages, with 101 illustrations, and 2 beautifully colored plates. Cloth, $2 25.

SENN (NICHOLAS). SURGICAL BACTERIOLOGY. In one hand-some octavo volume of 259 pages, with 13 plates, 9 of which are colored. Cloth, $1 75. *Just ready.*

SERIES OF CLINICAL MANUALS. A series of authoritative mono-graphs on important clinical subjects, in 12mo. volumes of about 550 pages, well illustrated. The following volumes are now ready: Ball on the Rectum and Anus, $2 25 ; Carter and Frost's Ophthalmic Surgery, $2 25 ; Hutchinson on Syphilis ($2 25) ; Marsh on Diseases of the Joints ($2) ; Morris on Surgical Diseases of the Kidney ($2 25); Owen on Surgical Diseases of Children ($2) ; Pick on Fractures and Dislocations ($2) ; Butlin on the Tongue ($3 50) ; Savage on In-sanity and Allied Neuroses ($2), and Treves on Intestinal Obstruc-tion ($2). The following are in press: Broadbent on the Pulse ; Lucas on Diseases of the Urethra.
For separate notices, see under various authors' names.

SIMON (W.) MANUAL OF CHEMISTRY. A Guide to Lectures and Laboratory work for Beginners in Chemistry. A Text-book specially adapted for Students of Pharmacy and Medicine. New (2d) edition. In one 8vo. volume of 480 pages, with 44 wood-cuts and 7 colored plates of deposits. Cloth, $3 25.

SKEY (FREDERIC C.) OPERATIVE SURGERY In one 8vo. vol. of 661 pages, with 81 woodcuts. Cloth, $3 25.

SLADE (D. D.) DIPHTHERIA; ITS NATURE AND TREATMENT. Second edition. In one royal 12mo. vol. pp. 158. Cloth, $1 25.

SMITH (EDWARD). CONSUMPTION; ITS EARLY AND REME- DIABLE STAGES. In one 8vo. vol. of 253 pp. Cloth, $2 25.

SMITH (J. LEWIS). A TREATISE ON THE DISEASES OF IN- FANCY AND CHILDHOOD. Sixth edition, revised and enlarged. In one large 8vo. volume of 867 pages, with 40 illustrations. Cloth, $4 50; leather, $5 50.

SMITH (STEPHEN). OPERATIVE SURGERY. New (second) and thoroughly revised edition. In one very handsome 8vo. volume, of 892 pages, with 1005 illustrations. Cloth, $4; leather, $5.

STILLÉ (ALFRED). CHOLERA, ITS ORIGIN, HISTORY, CAUSA- TION, SYMPTOMS, LESIONS, PREVENTION AND TREAT- MENT. In one handsome 12mo. volume of 163 pages, with a chart showing routes of previous epidemics. Cloth, $1 25.

STILLÉ (ALFRED). THERAPEUTICS AND MATERIA MEDICA. Fourth revised edition. In two handsome octavo volumes of 1936 pages. Cloth, $10; leather, $12; very handsome half Russia, $13.

STILLÉ (ALFRED) AND MAISCH (JOHN M) THE NATIONAL DISPENSATORY: Containing the Natural History, Chemistry, Pharmacy. Actions and Uses of Medicines. Including those rec- ognized in the latest Pharmacopœias of the United States, Great Britain and Germany, with numerous references to the French Codex. New (fourth) edition, revised and enlarged with an Appen- dix. In one magnificent imperial octavo volume of 1794 pages, with 311 accurate engravings on wood. Cloth, $7 25; leather, raised bands, $8; very handsome half Russia, raised bands and open back, $9 Also, furnished with Ready Reference Thumb letter Index for $1 in addition to price in any of the above styles of binding.

STIMSON (LEWIS A.) A TREATISE ON FRACTURES AND DISLOCATIONS. In two handsome octavo volumes. Vol. I., Frac- tures, 582 pages, 360 beautiful illustrations. Vol. II., Dislocations, 540 pp., 163 illustrations. Complete work, cloth, $5 50; leather, $7 50. Either volume separately, cloth, $3; leather, $4.

—— A MANUAL OF OPERATIVE SURGERY. New edition. In one royal 12mo. volume of 503 pages, with 342 illustrations. Cloth, $2 50.

STUDENTS' SERIES OF MANUALS. A series of fifteen Manuals by eminent teachers or examiners. The volumes are pocket-size 12mos of from 300–540 pages, profusely illustrated, and bound in red limp cloth. The following volumes may now be announced: Bruce's Materia Medica and Therapeutics (fourth edition), $1 50; Treves' Manual of Surgery (monographs by 33 leading surgeons), 3 volumes, each $2 00; Bell's Comparative Physiology and Anatomy, $2 00; Robertson's Physiological Physics, $2 00; Gould's Surgical Diagnosis, $2 00; Klein's Elements of Histology (3d edition), $1 50; Pepper's Surgical Pathology, $2 00; Treves' Surgical Ap- plied Anatomy, $2 00; Power's Human Physiology, second edition, $1 50; Ralfe's Clinical Chemistry, $1 50; and Clarke and Lock- wood's Dissector's Manual, $1 50. The following is in press: Pep- per's Forensic Medicine. For separate notices, see under various authors' names.

STURGES (OCTAVIUS). AN INTRODUCTION TO THE STUDY OF CLINICAL MEDICINE. In one 12mo. vol. Cloth, $1 25.

TAIT (LAWSON). DISEASES OF WOMEN AND ABDOMINAL SURGERY. Handsome octavo volume, 600 pages, fully illustrated. *Preparing.*

TANNER (THOMAS HAWKES). A MANUAL OF CLINICAL MEDI-CINE AND PHYSICAL DIAGNOSIS. Third American from the second revised English edition. Edited by Tilbury Fox, M.D. In one handsome 12mo. volume of 362 pp., with illus. Cloth, $1 50.

—— ON THE SIGNS AND DISEASES OF PREGNANCY. From the second English edition. In one 8vo. volume of 490 pages, with four colored plates and numerous woodcuts. Cloth, $4 25.

TAYLOR (ALFRED S.) MEDICAL JURISPRUDENCE. Eighth American from tenth English edition, specially revised by the Author. Edited by John J. Reese, M.D. In one large octavo volume of 937 pages, with 70 illustrations. Cloth, $5; leather, $6.

—— ON POISONS IN RELATION TO MEDICINE AND MEDICAL JURISPRUDENCE. Third American from the third London edition. In one octavo volume of 788 pages, with 104 illustrations. Cloth, $5 50; leather, $6 50.

—— THE PRINCIPLES AND PRACTICE OF MEDICAL JURIS-PRUDENCE. Third ed. In two handsome 8vo. vols. of 1416 pp., with 188 illustrations. Cloth, $10; leather, $12.

TAYLOR (ROBERT W.). A CLINICAL ATLAS OF VENEREAL AND SKIN DISEASES. Including Diagnosis, Prognosis, and Treatment. In eight large folio parts, measuring 14 x 18 inches, and comprising 213 beautiful figures on 58 full-page chromo-lithographic plates, 85 fine engravings, and 425 pages of text. Complete work, just ready. Price per part, $2 50. Bound in one volume, half Russia, $27; half Turkey Morocco, $28. *For sale by subscription only* Address the Publishers. Specimen plates by mail on receipt of ten cents.

—— THE PATHOLOGY AND TREATMENT OF VENEREAL DIS-EASES. Being the sixth edition of Bumstead and Taylor. In one very handsome 8vo. volume of about 900 pages, with about 150 engravings as well as chromo-lithographic plates. *Preparing.*

THOMAS (T. GAILLARD). A PRACTICAL TREATISE ON THE DISEASES OF WOMEN. Fifth edition, thoroughly revised and rewritten. In one large and handsome octavo volume of 810 pages, with 266 illustrations. Cloth, $5; leather, $6; very handsome half Russia, $6 50.

THOMPSON (SIR HENRY). CLINICAL LECTURES ON DISEASES OF THE URINARY ORGANS. Second and revised edition. In one octavo volume of 203 pages, with illustrations. Cloth, $2 25.

THOMPSON (SIR HENRY). THE PATHOLOGY AND TREAT-MENT OF STRICTURE OF THE URETHRA AND URINARY FISTULÆ. From the third English edition. In one octavo volume of 359 pages, with illustrations. Cloth, $3 50.

TIDY (CHARLES MEYMOTT). LEGAL MEDICINE. Volumes I. and II. Two imperial octavo volumes containing 1193 pages, with 2 colored plates. Per volume, cloth, $6; leather, $7.

TODD (ROBERT BENTLEY). CLINICAL LECTURES ON CERTAIN ACUTE DISEASES. In one 8vo. vol. of 320 pp., cloth, $2 50.

TREVES (FREDERICK). A MANUAL OF SURGERY. In Treatises by 33 leading surgeons. Three 12mo. volumes, containing 1866 pages, with 213 engravings. Price per set, $6. See *Students Series of Manuals*, p. 14.

—— SURGICAL APPLIED ANATOMY. In one 12mo. volume of 540 pages, with 61 illustrations. Cloth, $2 00. See *Students' Series of Manuals*, page 14.

—— INTESTINAL OBSTRUCTION. In one 12mo. volume of 522 pages, with 60 illustrations. Cloth, $2 00. See *Series of Clinical Manuals*, p. 13.

TUKE (DANIEL HACK). THE INFLUENCE OF THE MIND UPON THE BODY. Second edition. In one handsome 8vo. vol. of 467 pages, with 2 colored plates. Cloth, $3.

VAUGHAN (VICTOR C.), and NOVY (FRED'K G.) PTOMAINES AND LEUCOMAINES, OR PUTREFACTIVE AND PHYSIOLOGICAL ALKALOIDS. In one handsome 12mo. volume of 311 pages. Cloth, $1 75.

VISITING LIST. THE MEDICAL NEWS VISITING LIST for 1890. Thoroughly revised. 48 pages of indispensable data, and 176 pages of conveniently ruled and classified blanks for records. Pocket, pencil, catheter scale, and erasable tablet. Three styles: Weekly (dated, for 30 patients); Monthly (undated), and Perpetual (undated) Each in one vol., price, $1 25. With thumb-letter index for quick use, 25 cents extra. Special rates to advance-paying subscribers to The Medical News or The American Journal, or both. See p. 1.

WALSHE (W. H.) PRACTICAL TREATISE ON THE DISEASES OF THE HEART AND GREAT VESSELS. 3d American from the 3d revised London edition. In one 8vo. vol. of 420 pages. Cloth, $3.

WATSON (THOMAS). LECTURES ON THE PRINCIPLES AND PRACTICE OF PHYSIC. A new American from the fifth and enlarged English edition, with additions by H. Hartshorne, M.D. In two large 8vo. vols. of 1840 pp., with 190 cuts. Clo., $9; lea., $11.

WELLS (J. SOELBERG). A TREATISE ON THE DISEASES OF THE EYE. New (fifth) edition, thoroughly revised. In one large and handsome octavo vol. of about 800 pages, with colored plates and about 300 woodcuts, as well as selections from the test-types of Jaeger and Snellen.

WEST (CHARLES). LECTURES ON THE DISEASES PECULIAR TO WOMEN. Third American from the third English edition. In one octavo volume of 543 pages. Cloth, $3 75; leather, $4 75.

—— ON SOME DISORDERS OF THE NERVOUS SYSTEM IN CHILDHOOD. In one small 12mo. vol. of 127 pages. Cloth, $1.

WILLIAMS (CHARLES J. B. and C. T.) PULMONARY CONSUMPTION: ITS NATURE, VARIETIES AND TREATMENT. In one octavo volume of 303 pages. Cloth, $2 50.

WILSON (ERASMUS). A SYSTEM OF HUMAN ANATOMY. A new and revised American from the last English edition. Illustrated with 397 engravings on wood. In one handsome octavo volume of 616 pages. Cloth, $4; leather, $5.

—— THE STUDENT'S BOOK OF CUTANEOUS MEDICINE. In one handsome royal 12mo. vol. Cloth, $3 50.

WINCKEL ON PATHOLOGY AND TREATMENT OF CHILDBED. With additions by the Author. Translated by James R. Chadwick, A.M., M.D. In one handsome 8vo. vol. of 484 pages. Cloth, $4.

WÖHLER'S OUTLINES OF ORGANIC CHEMISTRY. Translated from the 8th German edition, by Ira Remsen, M.D. In one 12mo. volume of 550 pages. Cloth, $3 00.

WOODHEAD (G. SIMS). PRACTICAL PATHOLOGY. A Manual for Students and Practitioners. In one beautiful octavo vol. of 497 pages, with 136 exquisitely colored illus. Cloth, $6.

YEAR-BOOK OF TREATMENT FOR 1889. A Comprehensive and Critical Review for Practitioners of Medicine. In contributions by 22 well-known medical writers. 12mo., 349 pp. Limp cloth, $1 25. For special rate with The Medical News and The American Journal of the Medical Sciences, see page 1.

YEAR-BOOK OF TREATMENT FOR 1887. Similar to above. 12mo., 341 pages. Limp cloth, $1 25.

YEAR-BOOK OF TREATMENT FOR 1886. Similar to above. 12mo., 320 pages. Limp cloth, $8 25.